SWEDISH SOCIAL DEMOCRACY AND EUROPEAN INTEGRATION

To my family

Swedish Social Democracy and European Integration

European Integration

The People's Home on the Market

NICHOLAS AYLOTT
Keele University, UK

Routledge
Taylor & Francis Group

LONDON AND NEW YORK

First published 1999 by Ashgate Publishing

Reissued 2018 by Routledge
2 Park Square, Milton Park, Abingdon, Oxon, OX14 4RN
711 Third Avenue, New York, NY 10017, USA

Routledge is an imprint of the Taylor & Francis Group, an informa business

Notice:
Product or corporate names may be trademarks or registered trademarks, and are used only for identification and explanation without intent to infringe.

Publisher's Note
The publisher has gone to great lengths to ensure the quality of this reprint but points out that some imperfections in the original copies may be apparent.

Disclaimer
The publisher has made every effort to trace copyright holders and welcomes correspondence from those they have been unable to contact.

A Library of Congress record exists under LC control number: 99073319

ISBN 13: 978-1-138-34556-0 (hbk)
ISBN 13: 978-1-138-34559-1 (pbk)
ISBN 13: 978-0-429-43780-9 (ebk)

Contents

List of figures

List of tables

Preface

This book started life as a doctoral thesis, although the manuscript has undergone considerable revision and updating since it was defended. It deals with what I consider to be one of the more fascinating questions about politics in the endlessly intriguing continent of Europe. What is it that makes a successful political party lose its internal unity over a particular issue? Each side of the disagreement suffers, because disunity at election time can often be damaging to the party's prospects, and this will probably weaken its ability to implement the preferences of either side of the argument. So when and how does the issue become so important to the protagonists that they are prepared to sacrifice their party's electoral interests in the cause of pushing it in their preferred direction? The Swedish Social Democratic Party's division over European integration offers an excellent case study of this political phenomenon, and I hope that the book sheds light on it.

I have also been attracted to this case because of the nature of the issue over which division occurred. Why the EU is so much more controversial in some countries than in others is, of course, an essential ingredient of the story told here. But no European, even disinterested political scientists, can ignore the normative dimension to the process. It profoundly affects the way we are governed, and this makes it interesting irrespective of all other analytical considerations.

Throughout the book, I have used the footnote system of referencing, as I see it as more convenient for the reader than having to look up endnotes on different pages, and because the Harvard system can clutter the main part of the text. I have included in the bibliography those newspaper and journal articles that constitute evidence in themselves for my arguments—editorials representing certain points of view, for example, or those penned by authors whose opinions are particularly relevant. Other articles, most of which simply supply facts or quotations, are acknowledged only in footnotes. Percentages in tables are usually rounded up, so totals do not always sum to exactly 100.

I have tried to translate names of organisations and institutions into English wherever possible, although, following the usual convention, acronyms usually remain in their original form. I therefore refer to the Social Democratic Party, but also to SAP (rather than SDP). The reader can assume that quotations taken from non-English-language sources are translated by me, for which I naturally take responsibility.

As regards terminology, I try to refer to places and organisations in the correct geographical and temporal senses. Scandinavia, then, refers to Sweden, Norway and Denmark; the Nordic countries include these three plus Finland and Iceland. The Treaty on European Union, which came into force in November 1993, created the European Union, of which the European Community is one pillar. I thus refer to the EC or Community in contexts before November 1993, and EU or Union thereafter. The terms chosen to describe certain political institutions and organisations required a more active editorial decision. For instance, *partistyrelse* is often translated by Swedes as party board; but, following the standard terminology established by Katz and Mair, I refer to it as national executive. For the same reason, *verkställande utskott* I translate as executive committee. By party leadership, I refer chiefly to the seven full members of the Social Democratic Executive Committee. The names of most groups within the Swedish labour movement are translated according to its English-language literature.

Concerning more common English expressions, grass-roots and rank-and-file I use interchangeably. The same applies to post-materialist and non-materialist; theoretical differences between the two positions could doubtless be proposed and justified, but such an enterprise is beyond the scope of this study. The same applies to Eurosceptics, anti-accessionists and No-sayers, all of which refer to people who oppose membership of the EU. (Anti-European, on the other hand, is a term I try to avoid in this context, as I think it connotes something different.) Indeed, I consider Euroscepticism, a term originally attached to British Conservative MPs who were hostile to further integration in the EC/EU, to be now sufficiently established in discourse on the topic as to be appropriate for generic use in a study such as this one.

A version of chapter 6 appeared as "Between Europe and Unity: The Case of the Swedish Social Democrats", *West European Politics* vol. 20, no. 2, 1997.

Acknowledgements

My thanks to those who have helped me in my research could run on for many pages, but I will keep it more concise than perhaps it ought to be. The first acknowledgement must go to Leeds Metropolitan University for providing the studentship and facilities that allowed me to complete the first three years of the project. Thanks must go especially to my doctoral supervisor, David Arter, who first fired my interest in things Nordic; for that I am especially grateful. My colleagues in the Department of Politics at Keele University, where I worked after these first years, not only tolerated the continuing distraction that my doctorate represented, but also actively supported my engagement in it. Indeed, the intellectual climate in the department, and my colleagues' general friendliness and advice, did a great deal to make this a better piece of work than it would otherwise have been.

The department also funded my attendance at the Essex Summer School in Research Methods in 1996, which was of considerable benefit to the theoretical approach and the data analysis employed in my work. Discussion with various members of the Department of Business Administration and Social Science at Luleå University of Technology, where I spent a two-month exchange visit in the spring and autumn of 1997, was also very useful. Special thanks go to Luleå's Stefan Ekenberg and Nils-Gustav Lundgren. Involvement in a project on Social Democratic Parties and EMU, run by Ton Notermans and ARENA (Advanced Research on the Europeanisation of the Nation State), which is funded by the Norwegian Research Council, was a great help in my focusing the issues surrounding the EU single currency and the question of Sweden's participation.

The Social Democratic Party itself could hardly have been more accommodating. In particular, Eva Olofsson, Conny Fredriksson, the party's former international secretary, and Jerry Wiklund did so much to arrange my interviews up and down Sweden, to administer my questionnaire survey and to offer me their candid views on various issues. My data would have been infinitely harder to compile without their generous co-operation. Svante Fregert and Pontus Hedqvist were and are constantly helpful in shaping my ideas on a range of topics. Draft chapters owe a lot to the suggestions of various friends and colleagues, including Bob Leach, Anders Widfeldt, Torbjörn Bergman and Andrea Chandler. The advice and encouragement that I enjoyed from the last of these was of special help, and I value it enormously. Thanks too to the friends, relations and colleagues who helped with proof-reading, and particularly to Michael Aylott for his diligence in checking the quantitative data.

Finally, and above all, thanks go to my whole family, and my parents in particular. Without their unstinting support, my research would probably never have started at all, and would certainly never have reached the stage that it has.

List of acronyms and abbreviations

c	Swedish Centre Party
CAP	common agricultural policy
DNA	Norwegian Labour Party
EC	European Community
ECB	European Central Bank
ECJ	European Court of Justice
Ecofin	Council of Economic and Finance Ministers
EEA	European Economic Area
EFO model	Edgren–Faxén–Ohdner model of wage-formation
EFTA	European Free Trade Association
EMS	European Monetary System
EMU	economic and monetary union
ERM	exchange-rate mechanism
EU	European Union
fp	Swedish Liberal Party
kd	Swedish Christian Democrats
LO	Swedish Confederation of Trade Unions
m	Swedish Moderate Party
mp	Swedish Green Party
MEP	member of the European Parliament
MP	member of parliament
NATO	North Atlantic Treaty Organisation
nyd	Swedish New Democracy party
OEEC	Organisation for European Economic Co-operation
OECD	Organisation for Economic Co-operation and Development
s	Swedish Social Democratic Party
SAP	Swedish Social Democratic Party
SD	Danish Social Democratic Party
SEA	Single European Act
SFIO	French Section of the Workers' International
SPD	German Social Democratic Party

SSU	Swedish Social Democratic Youth
v	Swedish Left Party
WFTA	wide free-trade area

1 Introduction: Sweden, Social Democracy and European integration

On November 13th 1994 Swedish voters decided in a referendum that their country should join the European Union, accepting the terms of accession agreed by their government eight months before. The referendum result was close, however: 52.2 per cent said Yes to Swedish membership, 46.9 per cent said No. The winning margin was just 295,000 votes. The issues of European integration and Sweden's role in it had become highly divisive in the country's political life. The fault lines in this division were and are varied, sometimes running parallel, sometimes cutting across each other. But they involved many of Sweden's institutions, and perhaps none more so than its political parties. In some cases, it pitted them against each other. The Liberals and the conservative Moderates were fairly united in favour of membership, the Left and Green parties even more solidly against. But three, the Centre, the Christian Democratic and the Social Democratic parties, were badly divided over whether Sweden should or should not join the Union (see table 1.1). The Social Democratic Party (*Socialdemokratiska Arbetarepartiet*, SAP) is Sweden's largest. It has won an average of 44.1 per cent in national elections since 1960, well over double that of the country's second-biggest party, and was in government for a remarkable 60 of the 77 years up to 1998. The implications of the Social Democrats' internal division are particularly far-reaching, for the party itself, for Sweden and, indeed, for the European left's approach to the issue of economic and political integration. It is the subject of this study.

During the 20th century, much of it spent under Social Democratic rule, Sweden had established a glowing international reputation for the success of its economy, the humanity of its welfare policies and the righteousness of the international causes to which it committed itself. Indeed, left-leaning academics from elsewhere, notably in the English-speaking world, hailed the success of the so-called Swedish model. In 1989 a Canadian, Henry Milner, described Sweden as "social democracy in practice".[1] An American academic, Tim Tilton, believed by the beginning of the 1990s that "it is clear that Sweden's Social Democrats are not idly boasting when they claim to have built a social democracy."[2] Such admiration engendered among Swedes, particularly those sympathetic to SAP, a singular pride in their country. It was rich, highly civilised, and in world affairs it became something of a crusader for the rights of small countries. Its neutrality between the superpowers became increasingly activist as the cold war progressed, and Sweden's leaders felt able to criticise America (over Vietnam, for instance) and the Soviet Union (over Afghanistan) equally. Indeed, Swedish nationalism evolved partly as a self-perceived contrast from the superpowers. Sweden was not big and impersonal, but small and accessible. It was not a dictatorship, but scrupulously democratic, with very high rates of popular electoral participation (average turnout in national elections since the war is an impressive 86.3 per cent). It was not imperialist, but anti-imperialist. It was in favour of free trade, yet its form of capitalism was not ruthless or cut-throat, but inclusive and egalitarian. Its women and ethnic minorities received very favourable treatment. Fundamental to this self-perception was the role of the Social Democrats. Their long years in government did much to foster the "welfare nationalism" that marked the Swedish character.

By the end of the 1980s, however, the pillars on which the success of the Swedish model rested had begun to crumble. The most obvious factor in this was the transformation of the international situation brought about by the collapse of communist rule in Central and Eastern Europe and the end of the cold war. To be sure, these events were as welcomed in Sweden as they were elsewhere. Yet for some European countries—the former communist ones, of course, but also others, such as Sweden's Nordic neighbour, Finland—the demise of Soviet power offered almost nothing but new and attractive opportunities. "By contrast," according to two commentators, "in Sweden, the end of the cold war era removed the *supporting* rather than the restraining structures of its identity as a nation

---

1 Henry Milner, *Sweden: Social Democracy in Practice* (Oxford, Oxford University Press, 1989).
2 Tim Tilton, *The Political Theory of Swedish Social Democracy* (Oxford, Clarendon Press, 1991), p.5.

Table 1.1 Party sympathy and voters' choice in Sweden's EU referendum, November 1994

	Voting in parliamentary election, Sept. 1994	Voting in referendum	
	%	Yes %	No %
Moderates	22.4	86	13
Liberals	7.2	79	20
Christian Democrats	4.1	54	44
Social Democrats	45.3	46	53
Centre Party	7.7	44	54
Greens	5.0	20	79
Left Party	6.2	13	85

Note: Spoilt ballot papers and other parties ignored.
Source: Holmberg, "Partierna gjorde så gott de kunde" (1996), p.226.

state."[3] Finnish neutrality had been essentially a strategy for survival. Sweden, on the other hand, had exploited its neutrality to pursue foreign policy independently of the superpowers and their allies. As the relevance of neutrality diminished, Sweden's special role in the world became less clear. After all, many asked who was there now to be neutral between? Even more debilitating was the steep decline of the Swedish economy. By the early 1990s the proportion of the workforce unemployed, long below 5 per cent, had risen to double figures. One political economist declared: "Though hardly extraordinary by comparative standards, these are staggering, almost unbelievable, numbers from a Swedish perspective."[4]

A growing, high-employment economy had underpinned the Social Democrats' remarkable electoral success, and created an image that had been sharpened by active neutrality in international affairs. By the 1990s the economy was in crisis and neutrality seemed irrelevant.

Sweden's new situation was epitomised by the change in its relationship with the European Community. Since the EC's inception in the

---

3 Lars Svåsand and Ulf Lindström, "Scandinavian Political Parties and the European Union", in John Gaffney (ed.), *Political Parties and the European Union* (London, Routledge, 1996), p.210. Emphasis in original.
4 Jonas Pontusson, "Sweden: After the Golden Age", in Perry Anderson and Patrick Camiller (eds), *Mapping the West European Left* (London, Verso, 1994), p.23.

1950s, Swedes had grown used to standing outside it. Certainly, there were Swedes, mainly on the centre-right but also within SAP, who advocated their country's accession from the beginning. But most had been more or less content, or at least resigned, to adhere to the official Swedish position that neutrality rendered impossible joining an organisation whose members were (until 1973) all also, through NATO, protagonists in the cold war. At the same time, among many in SAP there was a veiled sense of Sweden's having little to gain or even to learn from the Community. During the 1960s some leading Social Democrats bluntly disdained the prospect of accession on the grounds that EC countries had "a more primitive social organisation than our own".[5] Though rarely stated so explicitly, SAP's leaders had long sought to exploit politically the apparent superiority of the Swedish model; the country was contrasted with the raw, free-market capitalism and social conservatism that supposedly dominated political and economic life in the Community. Then, in the space of a few months in late 1990, a Social Democratic government changed its stance completely and announced its intention to apply for membership of the EC.

This study addresses two main research questions. First, why did SAP's leadership change its policy towards the EU so abruptly? And, second, why was the party, whose unity and discipline had been such vital ingredients in its success during the 20th century, so divided by the leadership's decision to do so?

Both questions demand an examination of the dynamics of making policy in a social democratic party, an appreciation of Swedish political circumstances and a position on the nature of European integration. They are also important to political science for a number of reasons. First and foremost, this study is of political parties and the ways in which they behave. To what extent is it a party's ideology that determines its approach to a particular issue? Is its behaviour sociologically conditioned, in that the social location of its leaders, members and supporters sets parameters within which the party has to operate? Or are other, perhaps baser motives more influential? The party is, of course, an institution; and, like all institutions, it is ultimately composed of individual people. To what extent can party behaviour be explained in terms of the self-interest of these individuals? Is party behaviour a manifestation of a series of struggles, both within the party and with other parties, for the satisfaction of individuals' desires

---

5 Tord Ekström, Gunnar Myrdal and Roland Pålsson, *Vi och Vasteuropa. Uppfordran till eftertanke och debatt* (Stockholm, Rabén och Sjögren, 1962), p.164.

and goals? Such questions have been the subject of voluminous academic debate. The case of Swedish Social Democracy's trials with the issue of European integration, meanwhile, offers an excellent testing ground for competing theories of party behaviour. This is largely because our research questions identify real political puzzles. Why *did* SAP act as it did over Europe during the 1990s? Why *has* the party been so divided?

But our topic is also interesting for a number of more substantive political reasons. The response of parties to European integration raises questions about basic tenets of party politics in liberal democracies. Parties are often ascribed numerous functions by the literature dealing with them, in addition to perhaps their most important one, that of organising and constituting government. They structure the popular vote; they integrate and mobilise citizens; and they represent and aggregate different interests, coordinating the pursuit of both common and conflicting objectives.[6] In short, they are vital mechanisms for maintaining linkage between governors and governed. The crisis of the party has been heralded at various times; such pronouncements have generally proved unfounded or premature.[7] But European integration undoubtedly offers parties new challenges. How can meaningful linkage be maintained when binding political decisions are being made in the EU on the basis of intergovernmental negotiations? Accountability is a fundamental problem when the decisions to which national politicians contribute are the consequence of complex horse-trading with (at present) 14 other sets of politicians—not to mention assorted European commissioners and members of the European Parliament. On the other hand, it is questionable whether conventional parties could hope to construct in European institutions a new level of democratic accountability. Could parties ever reach across their national boundaries and present genuinely common policy platforms to a genuinely European electorate?[8] Should they attempt it, parties' ties to their national grassroots would surely be placed under great strain. The alternative, however, may be to live with a growing democratic deficit at the EU level.

---

[6] Peter Mair, "Introduction", in Peter Mair (ed.), *The West European Party System* (Oxford, Oxford University Press, 1991), p.1.

[7] Cf. Peter Mair, "Party Organizations: From Civil Society to the State", in Richard S. Katz and Peter Mair (eds), *How Parties Organize: Change and Adaptation in Party Organizations in Western Democracies* (London, Sage, 1994).

[8] Cf. John Gaffney, "Introduction: Political Parties and the European Union", in John Gaffney (ed.), *Political Parties and the European Union* (London, Routledge, 1996), p.17.

Moreover, our topic is relevant to the development of European integration itself. For one thing, the large majority of the Union's 15 member states currently have a social democratic party holding government office. The approaches that these parties take will obviously be a significant factor in determining how integration progresses. Second, SAP's case, although striking, is by no means the only example of a major political party riven by differences over the European issue. As we shall see in the following chapter, other Nordic parties are similarly troubled by such divisions, and both Britain's major parties have long been plagued by them. Most West European party systems were forged primarily by the contest for economic power and resources; when the advent of universal suffrage "froze" contemporary political configurations, it was this contest that formed the basis of the familiar left–right spectrum. Supranational integration, however, is very hard to fit into this spectrum. It offers something for (nearly) everyone: free trade and open markets for the right; resource redistribution and European-level reregulation for the left; handsome subsidies for agricultural interests; channels of cultural and political expression for national minorities. It is not obviously either a left-wing or a right-wing project, and so cannot be fitted neatly into existing party political cleavages. Thus, conflict over integration is as often found within parties as between them. Such internal division can cripple a party. If it is often in office, the implications for the governance of its country are clear.

These are dilemmas that face all parties, but they are perhaps especially acute for social democratic ones. They traditionally place great emphasis on, *inter alia*, two things. First, social democrats have been keen to use the power of the state for economic purposes—for which the institutions and public-policy instruments of the EU offer obvious possibilities. Second, they have stressed the need for effective, participatory democratic structures—for which the scale of European integration poses basic problems. It might be characterised as a conflict of *politics* against *democracy*.

The structure of the study

The rest of the book develops on the following lines. Chapter 2 places the case of Swedish Social Democracy in its political and research context. It briefly reviews some examples of West European social democratic parties that, despite initial scepticism, have come solidly to support further integration between the countries of the EU. A closer look, however, is warranted at SAP's closest relations, its sister parties in Denmark and Norway, where, as in Sweden, integration has been the cause of considerable difficulty. Another brief review is then made of some of the theories of party behaviour, particularly those designed especially for application to social democratic parties, that may be of help in answering our research question.

Chapter 3 assesses the political conditions that SAP was facing from around the beginning of the 1990s. All political actors act within constraints—social, political, legal, historical, personal—and the Social Democratic leadership was certainly no different. Neutrality was for many years a political fact of life in Sweden, one that had great—although perhaps exaggerated—significance for any Swedish government's European policy. The chapter further examines the ideological baggage that party elites were carrying, derived from decades of political success in the domestic arena. It addresses the notion of a "Swedish model", a term often used vaguely in a wide variety of social and political contexts, but which here is applied to a particular theory of political economy that the two wings of the labour movement, the Social Democratic Party and the Confederation of Trade Unions (LO), developed between them. To be sure, this model had its successes. But the problems it was facing by the 1980s are, as will become clear, an essential ingredient in the arguments concerning European policy that are proffered in this study. Finally, it considers internal party structures. How free was the party leadership to pursue the European policies, indeed any policies, that it preferred?

Did the dispute represent a conflict of economic interest, or a fundamental clash of ideology? The presence of such fault lines would scarcely be surprising in a party so broadly based that it has occasionally succeeded in attracting support from over half the Swedish electorate. But their relevance for the divide over Europe can only be gauged through qualitative and quantitative research. The results of such research is presented in chapter 4. Chapter 5 then details the circumstances, both internal and external to the party, in which during 1990 a Social Democratic government dramatically changed Sweden's policy towards the EC.

The following two chapters bring the study up to date. Chapter 6 recounts, given the existence of a split within SAP, the strategies adopted by its leaders to manage it and minimise the political damage that flowed from it. Analysis of these management strategies permits a judgment on whether they had the desired pacifying result, or whether in fact they may have inadvertently aggravated the party's internal division. Chapter 7 then revisits the methodology used to analyse the party leadership's volte-face over EC membership in 1990 and applies it to a more recent European dilemma, that concerning the realisation of economic and monetary union (EMU)—a single European currency. The Social Democratic elite's approach to this question has been rather different to the way it handled the question of EC membership itself at the beginning of the decade, and comparison of the two episodes allows some interesting observations to be made. Of course, the identity of the individuals in positions of power within the party changed in the meantime; but so did prevailing political and economic conditions.

Finally, chapter 8 draws some conclusions from the study, specifically for Swedish Social Democracy in Europe and more generally for the way political actors, especially those on the left, have approached the process of European integration, and how they might continue to do so in the early years of the new millennium.

2 Swedish Social Democracy in perspective: the political and research context

Social Democracy and European integration

West European social democratic parties have not always taken the positive attitude towards European integration that most take today. Germany's SPD was cool towards the establishment of the Council of Europe in 1949 and the European Coal and Steel Community in 1952, fearing that both would act as a distraction from its immediate priority of German reunification. In France, the SFIO was divided over the early steps towards integration, and until the mid-1980s a Eurosceptical tendency in its successor, the Socialist Party, expressed "a wide-ranging left-wing critique of the way the EC operates".[1] The now defunct Italian Socialist Party was unenthusiastic about Italian involvement until the 1960s. Large sections of Ireland and Britain's Labour parties were opposed to their countries' accession to the Community in 1973, and the latter advocated British withdrawal until as late as 1983. Greece's Socialists only became reconciled to membership after winning power for the first time in 1981, the year the country joined the EC. For the anti-integration elements within these parties, the Community was an obstruction to the political and economic reforms they hoped to see implemented. It was a rich man's club, an alliance of conservative governments whose aim, in freeing trade between them, was to enhance the power of international capitalism. In addition, the EC's decision-making structure was seen by some on the left as anti-democratic, with its technocratic Commission, its secretive Council of Ministers and its toothless, indirectly elected Assembly.

Various factors changed the social democratic approach to the EC. As we shall see, national circumstances greatly influence a party's policies.

---

1 Kevin Featherstone, *Socialist Parties and European Integration: A Comparative History* (Manchester, Manchester University Press, 1988), p.249.

9

Whatever his criticisms of the EU today, a Greek Socialist would find it difficult to eschew the 3-6 per cent of that country's GDP that emanates from the EU's coffers. Greece's Socialist government urged a Union "evolving towards the direction of deeper integration, with democratically structured and legitimate institutions" before the EU's intergovernmental conference in 1996.[2] In Germany, meanwhile, the historical legacy of the Nazi era discredited the concept of nationalism, so much so that to question adherence to the ideals of its supposed antidote, European integration, has become virtually taboo. In March 1996 the German Social Democrats' leader played a cautiously Eurosceptical card in regional elections, seeking to exploit German voters' fears about EMU. The SPD's reversals at the polls, and the criticism its leader received from some in his own party, indicated the depth of the country's consensus about support for the EU. Indeed, in the main, the Social Democrats are now at least as enthusiastic about integration as the Christian Democrats. Europe was barely mentioned in the 1998 election campaign that brought the Social Democratic candidate, Gerhard Schröder, the chancellorship.

In Britain, meanwhile, four consecutive election defeats between 1979 and 1992 forced a fundamental rethink in the Labour Party about European integration. The party's commitment to withdraw from the EC was dropped after Margaret Thatcher's second election victory in 1983, and a speech by the then president of the European Commission, Jacques Delors, to the Trades Union Congress in 1988 is widely credited with marking a watershed in the British labour movement's perception of European integration. With the Conservatives entrenched at the national level of policy-making, and a socialist leading the executive branch of the European level, the latter became relatively more attractive to those pursuing a social democratic agenda. Now Labour seems happy to present a broadly pro-integration image.

Moreover, the political climate has changed throughout the western world in the past 20 years. The advance of market-orientated politics in the 1980s was not confined to Britain and America; its effects were felt throughout Europe. The interest in promoting market solutions to the problem of "Eurosclerosis" was shared by such right-of-centre leaders as Jacques Chirac in France and Helmut Kohl in West Germany. Particularly after the failure of the attempt by the Socialist government of François

---

2 *Agence Europe* April 2nd 1996.

Mitterrand to reflate the French economy unilaterally in 1981-83, and in the light of the boom engineered by the Thatcher government in Britain during the mid-1980s, social democrats felt compelled to make doctrinal concessions to the turn of the political tide. No longer were moves to free trade between the EC's member states greeted with suspicion. Now

> Parties of the left see the European Union as a unit large enough to sustain its own economic, labour and trade policies in a way that no single nation could; and, as such, as a means for implementing effective policies in such areas as pollution control, worker consultation, labour market regulation and macroeconomic policy co-ordination.[3]

This growing, cross-party consensus on acceptance of the desirability of freer intra-Community trade culminated in the Single European Act (SEA) of 1985, which agreed the decision-making reforms deemed necessary to secure the completion of the EC's internal market by the end of 1992. The left's reconciliation with the Community was completed in the aftermath of the collapse of communism in Eastern Europe after 1989. Comprehensive state planning was finally discredited. An international market economy, regulated internationally, became social democracy's preferred goal.

Paradoxically, however, this conversion has been accompanied by an apparent waning of popular enthusiasm for the EU, a tendency that became most visible in Denmark. The People's Movement Against the EC/EU has run lists in the country's four direct elections to the European Parliament, and has consistently won around a fifth of the vote. Far more significant, however, was the Danish electorate's rejection, by a narrow margin, of the Treaty on European Union, agreed by the then 12 EU governments at Maastricht in December 1991. Had voters in France chosen not to approve the treaty, most observers believed that it would have fallen, whereas the tiny majority that said Yes was enough to maintain its passage. In contrast, its initial rejection in Denmark's referendum was not enough to derail it; Denmark was sufficiently small for other member states to be able persuade its government to ask the same question of its people a year later, in 1993, which drew a more positive response.[4] But Danish experi-

---

3 *The Economist*, "The left in Western Europe", June 11th 1994.

4 Opinion was divided over the exact nature of the question put to Danish voters in their second referendum on the Maastricht treaty. For some, the amendment worked out at the EU's Edinburgh summit in December 1992 gave Denmark significant exceptions from the original treaty; see, for example, Nikolaj Petersen, "'Game, Set and Match for Denmark':

ence was a pointer to the sort of difficulties Sweden would soon face, both in terms of the national electorate and its individual political parties.

Scandinavian social democratic parties and the problem of European integration

The referendums on the Maastricht treaty were the third and fourth that Denmark has held on European integration; the others were on EC membership in 1972 and on the Single European Act in 1986. The leadership of its Social Democratic Party (SD) has usually struck a cautiously pro-integration line, with the exception of the referendum on the Single Act, which—largely for tactical reasons, as it was in opposition at the time—it campaigned against.[5] But its voters have often been more sceptical. Only a small majority, 55 per cent, was persuaded to back the leadership's support for Danish accession to the EC in 1972.[6] In direct elections to the European Parliament, the Danish Social Democrats have suffered damaging defections by their usual supporters to Eurosceptics' electoral lists (see table 2.1); indeed, Worre estimates that around two-fifths either abstained or defected in the European elections of 1979 and 1984.[7]

In 1986 70 per cent of Social Democratic voters concurred with the party leadership's opposition to the Single Act (although the wider

---

Denmark and the EU After Edinburgh", in Teija Tiilikainen and Ib Damgaard Petersen, *The Nordic Countries and the European Community* (Copenhagen, Copenhagen Political Studies Press, 1993). Others believed that the agreement simply emphasised options to draw back from certain activities (such as monetary union, common defence and common citizenship) that Denmark already possessed; see, for instance, Edward Mortimer, "Same deal as before: the Danes did not win new concessions on Maastricht", *Financial Times* January 27th 1993. The ratification saga spawned a second Eurosceptical group in Denmark. The June Movement was formed on the basis of opposition to Maastricht, and in the 1994 election to the European Parliament it won 15.2 per cent of the vote, to 10.3 per cent won by the People's Movement.

[5] However, several leading figures in the party defied the leadership's line and supported the treaty. Torben Worre, "Denmark at the Crossroads: The Danish Referendum of 28 February 1986 on the EC Reform Package", *Journal of Common Market Studies* vol. 26, no. 4, 1988, p.371.

[6] Jens Henrik Haahr, *Looking to Europe: The EC Policies of the British Labour Party and the Danish Social Democrats* (Aarhus, Aarhus University Press, 1993), p.179.

[7] Torben Worre, "The Danish Euro-Party System", *Scandinavian Political Studies* vol. 10, no. 1, 1987, p.87.

electorate voted to approve it).[8] But its recommendation that Maastricht be accepted was emphatically rejected by most Social Democratic supporters: just 33 per cent voted Yes in the first referendum on the treaty in June 1992 (in which it was rejected), and only 50 per cent did so even in the second (in which it was approved)[9]—before which several European leaders had hinted that a second No might mean Denmark's *de facto* expulsion from the Community.

Norway has twice voted on membership of the EC/EU, in 1972 and 1994. Its first referendum arguably inflicted even graver damage on the Norwegian Labour Party (DNA) than that suffered by SD. The Labour leader and prime minister, Trygve Bratteli, had been in favour of membership, and he eventually mobilised the party's organisation to campaign for a Yes. In the event, however, 53.5 per cent of Norwegians who voted rejected the proposal, including a third of DNA voters.[10] The leadership's failure to secure the result it wanted had a disastrous effect on the party's electoral fortunes. Bratteli resigned as prime minister, Labour lost power and in the national election a year later the party's share of the vote fell from the 46 per cent to 35 per cent. In 1969 Labour claimed half of first-time voters; in 1973 the proportion was a quarter. Several months before the election, a left-wing section of the party defected to form a Socialist Electoral Alliance with the Socialist People's Party and the Communists; half the Alliance's 11 per cent return comprised former Labour voters. Just as damagingly, nearly a third of the 5 per cent achieved by a new, populist, anti-tax and pro-EC party, led by Anders Lange, came from DNA's camp.[11]

---

[8] Ole Tonsgaard, "Folkeafstemningen om EF-pakken", *Dansk Udenrigspolitisk Årbog 1986* (Copenhagen, DJØF and DUPI, 1987). Cited in Haahr (1993), p.211.

[9] Torben Worre, "First No, then Yes: The Danish Referendums on the Maastricht Treaty 1992 and 1993", *Journal of Common Market Studies* vol. 33, no. 2, 1995, p.246.

[10] Henry Valen, "Norway: 'No' to EEC", *Scandinavian Political Studies* vol. 8, 1973, p.218.

[11] See Hillary Allen, *Norway and Europe in the 1970s* (Oslo, Universitetsforlaget, 1979), pp.195-99.

Table 2.1 Social democratic parties, electoral performance and the issue of European integration in Scandinavia

	Danish Social Democrats			Norwegian Labour			Swedish Social Democrats		
	Nat	EP	Ref	Nat	EP	Ref	Nat	EP	Ref
1971	37								
1972			50 (63)*			65 (47)*			
1973	26			35			43*		
1974									
1975	30						43		
1976									
1977	37			42*					
1978									
1979	38*	22					43		
1980									
1981	33			37*					
1982							46		
1983									
1984	32	20							
1985				41			45*		

Year								
1986			30					
1987	29							
1988	30	23						
1989				34*				43*
1990	37							
1991								38*
1992			33 (49)					
1993			50 (57)*	37				
1994	35*	16			64 (48)*	45		
1995							28*	
1996								
1997				35*				
1998	36*		(55)*			36*		
1999		17*				26*		46 (52)*

Key: Nat = national election. Ref = referendum on European integration: EC/EU membership (Denmark; 1972; Norway, 1972 and 1994; Sweden, 1994); the Single European Act (Denmark, 1986); and the Treaty on European Union (Denmark, 1992 and 1993). EP = direct elections to the European Parliament. * = election/referendum fought with party in government.

Notes: All figures rounded to nearest integer. Figures for referendums and European elections denote proportion of social democratic voters estimated to have voted Yes (with the Yes vote among the wider electorate in brackets).

Sources: Allen (1979); Bjørklund (1996); Haahr (1992); Miljan (1977); Petersson (1994); Valen (1973); Worre (1995); election returns.

Of course, there have been basic differences between the situations in Denmark, Norway and Sweden since 1972. The national backgrounds to the European issues were markedly different. Some argue that Norwegian nationalism was aggravated by the EC's use of the word union to describe its goals: Norway had been coerced into unions with Denmark (until 1814) and Sweden (until independence in 1905), and had acquired sovereign status only relatively recently. Economic situations also differed. For Denmark, the need to preserve access to the British food market was over-riding, so much so that "Denmark did not become a member of the [EC] for idealistic or political reasons but because pragmatic economic considerations left no alternative."[12] Norwegians, meanwhile, with their large oil resources, were suspicious of the EC's plans for a common energy policy. Moreover, the role of the primary sector—the most significant single area of economic life subject to the Community's competence—was greater in the Norwegian No lobby than in Sweden's, where farmers and fishermen were ambivalent about accession, or in Denmark, where they favoured EC membership. The problems of adapting to the EC's developing fishing regime led to Norway's fisheries minister resigning when the country's terms of accession were concluded in 1972. In contrast, Sweden's terms of entry received a broad welcome from across the political spectrum. Perhaps as important as anything were the respective economic situations. For instance, Norway's economy was booming in 1972. Sweden's decidedly was not in 1994.

All these factors meant that Norwegian opposition to accession in 1972 was considerably stronger than it was in Denmark in the 1970s, 1980s and 1990s, and in Sweden in the 1990s. All the Swedish non-socialist, or bourgeois, parties' leaderships were united in favour of a Yes. The Danish bourgeois parties have mostly been in favour of integration, the Liberals and Conservatives especially. By contrast, in 1972 the Norwegian Centre, Liberal and Christian People's parties all urged voters to say No to EC membership. The anti-accession camp in Norway had the strong financial support of both the farmers' and fishermen's organisations and, crucially, the sympathy of a significant part of the press. In Sweden, neither primary-sector organisation opposed membership, and nor did all but a handful of regional newspapers. In Denmark, the primary sector was and

---

[12] Toivo Miljan, *The Reluctant Europeans: The Attitudes of the Nordic Countries towards European Integration* (London, C. Hurst and Co., 1977), p.279.

continues to be aware of its vital interest in remaining within the EU's agricultural regime.

Finally, the international contexts in which each referendum took place were very different. Although there was broad consensus in all three countries about the bases of their security policies—the minorities that supported Norwegian or Danish withdrawal from NATO in 1972-93, and abandonment of Swedish nonalignment in 1994, were similarly small—the Norwegian and Danish debates were not comparably affected by a recent event in the way that Sweden's was by the end of the cold war.[13] If the demise of bloc confrontation in Europe did not in itself present a positive reason for Swedish accession to the Union, it undoubtedly removed the major impediment. In addition, concerning the timing of the referendums, the so-called domino effect did not work to the advantage of Norwegian supporters of accession in 1972 in the way that it did for those in Sweden in 1994. In the latter case, the four applicants' referendums were arranged so that those most likely to produce positive results would be held first and those least likely last, as the four governments hoped that the electorates in the least enthusiastic states might be more easily dissuaded from voting No by the prospect of their country's isolation outside the EU.[14] Pro-accession campaigners in Norway, which voted last in the round of referendums in 1994, had no such advantage in 1972. Then, as a sop to the party's left wing, Denmark's Social Democratic government set its referendum after the Norwegian vote, despite the fact that a Yes to membership was generally seen as more likely in Denmark than in Norway.

For all these reasons, comparisons between the plights of the three Scandinavian social democratic parties over European integration should not be taken too far. Yet there were similarities in the respective situations—particularly between those involving DNA in 1972 and SAP in 1994. Both parties found that the strength of their anti-accession elements was for some time obscured by external factors: neither party's No side wished to provoke an open split while the possibility remained that membership negotiations might yet founder. Like SD in 1986 and 1992, both

---

[13] The Danish votes on the Maastricht treaty took place, like Sweden's on accession, after the end of the cold war and the reunification of Germany. But, as a NATO member, such developments did not affect the very foundation of Denmark's security policy in the way they did neutral Sweden's.

[14] See Detlef Jahn and Ann-Sofie Storsved, "Legitimacy through Referendum? The Nearly Successful Domino-Strategy of the EU-Referendums in Austria, Finland, Sweden and Norway", *West European Politics* vol. 18, no. 4, 1995, pp.18-37.

parties found their disagreements over the EU easier to handle while they were in opposition (at least before the conclusion of their country's membership negotiations); without the responsibility of government, the leaderships of DNA and SAP could both be sufficiently equivocal about backing membership as to dissuade all strands of opinion from openly challenging its authority. When the opposition to EC/EU membership did emerge, it comprised in large part grass-roots opinion; in neither party was there even a semi-organised faction that aspired to use the European issue to advance its power within the party. Above all, both leaderships had the fundamental task of juggling two conflicting objectives: maintaining the party unity necessary to win an election, and yet supporting accession with enough vigour to secure a Yes in the referendum on accession. SAP, then, did not need to look far from Sweden to see the dangers that the European issue held.

Voting patterns in the Swedish referendum

In some ways, the referendum was something of a success for the Swedish political system—by the standards of previous referendums, at least. Turnout was 83.3 per cent, which was 3.5 per cent lower than in the parliamentary election two months previously, but which represented the highest-ever turnout in a Swedish national referendum. Of first-time voters, one in five (19.7 per cent) abstained; but over a third (33.7 per cent) had done so in Sweden's previous referendum, on nuclear power in 1980.[15] Whereas after the referendum in 1980 53 per cent believed that the Swedish parliament, the Riksdag, and the government ought to have taken the decision, the figure in 1994 was just 22 per cent.[16] Indeed, the referendum has something of a dubious pedigree in the Swedish political system.[17] In

---

15 Martin Bennulf and Per Hedberg, "Passiv eller activ väljare?", in Mikael Gilljam and Sören Holmberg (eds), *Ett knappt ja till EU. Väljarna och folkomröstning 1994* (Stockholm, Norstedts juridik, 1996), p.101.

16 Mikael Gilljam, "Den direkta demokratin", in Mikael Gilljam and Sören Holmberg (eds), *Ett knappt ja till EU. Väljarna och folkomröstning 1994* (Stockholm, Norstedts juridik, 1996), pp.20-22.

17 All referendums held in Sweden have been consultative, invoked by a simple majority in the Riksdag, although since 1980 there has been provision for a popular vote that can reject proposed constitutional change (which must be proposed by at least a tenth of MPs and supported by at least a third).

1955, for example, voters rejected, 82.5 per cent to 15.5 per cent, a proposal to change to driving on the right-hand side. Twelve years after the vote, the government nevertheless implemented the change.[18] The two other national referendums, on supplementary pensions in 1957 and on nuclear power in 1980, were both held largely because of tactical party-political calculations by Social Democratic governments.[19] In each of those, three choices were put to the voter. None won a majority, and policies ensued that voters had failed convincingly to endorse. By contrast, the referendum on the EU was rather different. With only two options, the choice was clear, and there could be no equivocation about the result, even if it was close.

Can research into voting behaviour in the Swedish referendum help us in our investigation into the division within the Social Democratic Party? Certainly, some relevant and interesting patterns can be observed. But in many ways these leave us with as many questions as answers. Above all, it is not clear why such patterns should form in response to European integration; the connections between the EU, on one hand, and the interests of specific social and economic groups, on the other, is not obvious to the outside observer.

The issues in the referendum campaign, if not the arguments, were reasonably settled. For the Swedish electorate, economic factors were the most important. Voters questioned both before and after the referendum most often named the economy as their primary motivation. After that, Yes voters seemed to have been persuaded significantly by the argument—propounded most explicitly by the Moderate leader, Carl Bildt—that Sweden would be "isolated" outside the EU.[20] Similarly, the chance to influence the EU from the inside was attractive to them. Peace and security were also

---

[18] Partly because all the main parties approved changing the law on driving in spite of the referendum result, the event "engendered a lingering scepticism in the electorate about the future willingness of the political elite to abide by such outcomes". Olof Ruin, "Sweden: The Referendum as an Instrument for Defusing Political Issues", in Michael Gallagher and Pier Vincenzo Luigi (eds). *The Referendum Experience in Europe* (London, Macmillan, 1996), p.181.

[19] Ruin declares that referendums in Sweden are used, "as so often elsewhere, to solve a knotty parliamentary situation creating disunity either between cooperating parties or within a particular party". Ruin (1996), p.174.

[20] In a newspaper article a few days before the referendum, in which leading politicians were asked for their five best reasons for voting Yes or No, Bildt put "Co-operation is better than exclusion" in all five places.

Table 2.2 Voters' most often-named arguments for or against Swedish EU membership

inclined to vote Yes	%	voted Yes	%
economy	44	economy	38
Sweden isolated	36	Sweden isolated	30
influence EU	20	peace	27
employment	19	influence EU	22
peace	14	employment	20
inclined to vote No	%	voted No	%
economy	20	economy	22
consumer issues	20	democracy	21
open borders	17	EU's organisation	18
EU's organisation	17	open borders	17
democracy	15	consumer questions	15

Note: The inclination to vote refers to voters' intentions before the referendum.
Source: Oskarson (1996), "Väljarnas vågskålar", p.128.

factors that influenced pro-accession voters, although in quite a vague sense. The EU was seen as a "peace project", having contributed to the absence of war between its members. Just 7 per cent of all voters thought that Sweden's security would be enhanced by EU membership,[21] and only 30 per cent of No voters felt that it would be worsened if Sweden joined.[22] On the No side, meanwhile, a perceived lack of democracy in the Union was a powerful argument. There were also fears about opening Sweden to the rest of Europe, and exposing it to crime, drugs and possibly also inferior consumer goods (see table 2.2). As Oskarson puts it, "voters were relatively agreed on which political areas or questions tipped the scales for or against the EU, but Yes and No voters interpreted differently the consequences that membership would bring in these areas."[23]

---

[21] Maria Oskarson, "Väljarnas vågskålar", in Mikael Gilljam and Sören Holmberg (eds), *Ett knappt ja till EU. Väljarna och folkomröstning 1994* (Stockholm, Norstedts juridik, 1996), p.131.

[22] Rutger Lindahl, "En folkomröstning i fredens tecken", in Mikael Gilljam and Sören Holmberg (eds), *Ett knappt ja till EU. Väljarna och folkomröstning 1994* (Stockholm, Norstedts juridik, 1996), p.151.

[23] Oskarson (1996), "Väljarnas vågskålar", p.133.

Much the same can be said about the debate within SAP, with the same issues being utilised by each side within the party for different ends. Some Social Democrats insisted that Swedish membership would help to preserve the country's welfare state and combat unemployment. Ingvar Carlsson, the leader of the party and a strong supporter of membership, declared: "the Swedish model can, in my opinion, be safely recreated on the European level."[24] Others expressed a conviction that jobs could only be created, and indeed preserved, if Sweden stayed outside the Union. One of the foremost economists in "Social Democrats Against the EU", the group representing opponents of membership in the party, warned that the Union's plan for economic and monetary union "stands in direct conflict with what since the 1930s has been Social Democracy's chief goal, namely the maintenance of full employment".[25] Similarly, the profile of the division about Sweden's capacity to pursue desirable objectives was often hard to predict. For example, Social Democrats of all ages usually approved wholeheartedly of goals like promoting a clean environment or aiding the third world. But there seems little obvious reason why younger people should necessarily have been more likely to judge that Swedish *autonomy from the EU* would advance these aims more effectively, while older Social Democrats preferred to stress how Swedish *influence within the EU* would better achieve the same ends.

Socio-economic factors

There is some evidence that the social cleavages identified by Lipset and Rokkan[26] can be relevant to attitudes towards European integration. Hansen and Bjørklund compare the sociological analysis of voting in Norway's referendum on EU membership to those the country has held on several other issues, and find that in each case, the pattern of the periphery—defined not just geographically, but also in terms of linguistic and religious "counter-cultures"—voting against the wishes of the political centre is repeated with remarkable consistency, whatever the issue addressed by the

---

[24] Ingvar Carlsson, "Europasamarbetet: ett vänsterprojekt" in Rolf Edberg and Ranveig Jacobsen (eds), *På tröskeln till EU*, (Stockholm, Tidens förlag, 1994), pp.19-20.

[25] Sören Wibe, "EG och ekonomin", in Lotta Gröning (ed.), *Det nya riket? 24 kritiska röster om Europa-Unionen* (Stockholm, Tidens förlag, 1993), p.200.

[26] See Seymour Martin Lipset and Stein Rokkan, "Cleavage Structures, Party Systems, and Voter Alignment" (1967), reprinted in Peter Mair (ed.), *The West European Party System* (Oxford, Oxford University Press, 1991).

referendum.[27] Indeed, in the early 1990s the Norwegian Centre Party achieved considerable political success by recasting itself as the champion of these counter-cultures, and strongly opposed EU membership.

In Sweden, however, the significance of social cleavages was arguably less clear, although Oscarsson concludes that, by a short margin, social and economic location was the variable that best predicted voters' views on the EU.[28] Men, the elderly, the urban dwelling and southerners were all more likely to say Yes to membership than women, the young, those living in the country and northerners. The middle class, the better educated, the employed, and those working in the service and private sectors were more likely to be for accession than the working-class, the less well educated, the unemployed, and those working in the manufacturing and public sectors. The polyglot and those well-travelled abroad were readier to vote for joining the Union than the monoglot and those less well-travelled. Gilljam and Holmberg and their colleagues talk repeatedly of a centre–periphery divide in the Swedish electorate, using the term periphery in both a geographical sense and in a social and economic sense. The better off tended to vote Yes, the not so well off to vote No.

Naturally, this was bound to pose problems for SAP's leadership. The less privileged in Swedish society are its natural constituency. It is also especially strong in northern Sweden, away from the three big cities, Stockholm, Gothenburg and Malmö; yet voters here were significantly opposed to the pro-membership line that Social Democratic leaders offered (see table 2.3). One party activist from Sweden's northern-most county, interviewed by the present author, said that she was so isolated in that region in arguing for EU membership that sometimes she felt "like a bad Social Democrat".

In contrast to the survey led by Gilljam and Holmberg, the socio-economic data obtained by a questionnaire survey undertaken in the course of this study sampled not Social Democratic supporters but Social Democratic activists. It shows some comparable trends to those found among Social Democratic voters and the wider electorate. Male activists comprised nearly three-quarters of the respondents, but over four-fifths of those likely to vote Yes. Women, meanwhile, comprised just over a quarter

---

27 Tore Hansen and Tor Bjørklund, "The Narrow Escape—Norway's No to the European Union", paper presented to ECPR joint sessions of workshops, Oslo, April 1996.
28 Henrik Oscarsson, "EU-dimensionen", in Mikael Gilljam and Sören Holmberg (eds), *Ett knappt ja till EU. Väljarna och folkomröstning 1994* (Stockholm, Norstedts juridik, 1996).

of the respondents, but made up under a sixth of those likely to vote Yes. Those in work were slightly more likely to vote Yes, those unemployed slightly more likely to vote No. Finally, having a university education made a respondent slightly less likely to vote No (though this was quite a weak correlation). (See table 2.3.)

But how much do these figures really tell us? It is not difficult to suggest reasons why people without much experience of travelling abroad, or possessing many language skills, or even being less educated, would be less likely to support Sweden's joining the EU. People are naturally wary of the unknown, and those for whom other European countries were relatively unknown could be expected to resist their country's entry into a supranational union. But even where socio-economic factors can be seen as a powerful *predictor* of voting intention in the EU referendum, often this cannot be considered an *explanation* of voting intention.

The cleavages identified by Lipset and Rokkan were forged by real economic or political conflict, which may be still relevant today, albeit at a lower level of intensity. The distribution of resources between classes, for example, is at the heart of modern political struggle in Western Europe. There are numerous examples of peripheral cultures' demands for autonomy from central state authorities being reflected in party systems. From the literature on European integration, Moravcsik makes a similar point about political differences being associated with economic interest. If, say, a dominant position in a national market is threatened by the prospect of access to that market from foreign competitors, then opposition to integration is likely. If, on the other hand, producers are attracted by access to a foreign market, or if consumers welcome the possibility of wider choice between products, then such groups are likely to favour integration.[29]

---

29 Andrew Moravscik, "Preferences and Power in the European Community: A Liberal Intergovernmentalist Approach", *Journal of Common Market Studies* vol. 31, no. 4, 1993, p.485.

Table 2.3 SAP activists' and supporters' voting intention (June 1994) in EU referendum

	Supporters			Voting intention	Activists		
	Yes %	No %			Yes %	Uncertain %	No %
Male	58	41	Male		44	20	35
Female	46	53	Female		25	33	43
18-30	40	58	18-25		13	-	88
31-45	53	46	26-35		48	24	28
46-60	56	43	36-50		36	22	42
61-80	58	41	51-		43	26	42
Gainfully employed*	51	48	Working		42	23	35
Unemployed*	39	47	Unemployed		23	30	47
Private sector	55	44	Private sector		36	27	37
Public sector (local govt)	60	39	Public sector		43	26	42
Public sector (central govt)	46	53					

	University education	No university education
	40	40
	31	22
	29	39

University education	62	37
At least three years high school	55	44
Additional education	43	55
Basic education	44	55
Southern Sweden	60	39
Stockholm	60	39
Western Sweden	61	39
East-central Sweden	53	46
South-east Sweden	44	55
West-central Sweden	48	51
Northern Sweden	39	60

Notes: Exclusion of missing or non-applicable cases means totals may not equal 100. * Denotes categorisation with Centre Party and Christian Democratic supporters.
Sources: Data for supporters adapted from Gilljam (1996).

It is not hard to see, for instance, why Norwegian farmers and fishermen should have been opposed to EU membership: both would have had to forgo public subsidy that was even higher in Norway than in the Union, and sole control of national fishing waters would have had to be surrendered.[30] Lindström also sees further interest-based Euroscepticism emanating from the growth of the public sector. In particular, he argues that the growth of "public households"—those in which the state has a high degree of responsibility for employment of parents and care of children—created a large electoral constituency that has only an indirect reliance on economic growth. This constituency, he implies, is less likely than private-sector workers to see the benefits of European integration. The diversification of their natural supporters' political preferences has, Lindström believes, become a major problem for Danish and Swedish Social Democrats. Yet in Norway, where the nationalisation of work hitherto carried out in the home has gone furthest, Labour has had fewest problems. This is because of the country's extremely propitious economic circumstances: "Petroleum, fish, metals and the national tourist industry—all sources of wealth that cannot locate to another country."[31] These made Norwegians more likely to feel that they had no great need for the EC/EU, partly because oil wealth could sustain a large public sector.

But the relevance of socio-economic factors is not always so clear in the Swedish case. For example, many workers in Sweden's large public sector were sceptical about the Union.[32] But this was not because they feared new competition: most work in social services, particularly those provided by local government, and were hardly threatened by foreign producers. Rather, it seemed to have been caused by fear that the high tax-levels needed to sustain the public sector would not be tenable in the Union.[33] Yet pro-EU Social Democrats insisted, quite plausibly, that the Swedish public sector would actually come under greater pressure if the country remained *outside* the EU, as the tax burden would have to be re-

---

30 See Christine Ingebritsen. "Norwegian Political Economy and European Integration: Agricultural Power. Policy Legacies and EU Membership". *Conflict and Cooperation* vol. 30, no. 4, 1995.
31 Ulf Lindström. *Euro-Consent, Euro-Contract or Euro-Coercion? Scandinavian Social Democracy, the European Impasse and the Abolition of Things Political* (Oslo, Scandinavian University Press. 1994). p.46.
32 In June 1994 the congress of the Municipal Workers' Union voted to take no position on EU membership. and urged LO to do the same. which the confederation subsequently did.
33 Lindström (1994). chs 3-4.

duced to offer an added incentive for capital to invest there. In sum, then, certain socio-economic patterns can be observed among the Swedish electorate in general, and among Social Democratic voters and activists in particular. But we still require an explanation of these patterns. This is especially the case where the correlation between socio-economic circumstances and voting intention is weak. But even where it is stronger, often we still lack an explanation of *why* it should have influenced voting in the referendum.

Party behaviour

As Gaffney notes, while the literature on European integration and political parties is voluminous, "The literature on political parties in the Union, however, is minimal."[34] There are various theoretical approaches to studying the way political parties behave in relation to voters, to each other and to the institutions of the state. Ware classifies them as being founded on, respectively, sociological, institutional and competition-based explanations.[35] Naturally enough, all these approaches have their heuristic value. They each have very different starting points, but the three approaches are by no means wholly irreconcilable. The sociological approach has been most useful in identifying the social origins of political parties and, more indirectly, how this influences their contemporary behaviour. It has its most significant expression in the work of Lipset and Rokkan. The institutional approach deals not so much with parties' origins but with their internal organisation and legal and political environment. But it is the third approach, the competition-based approach associated with Downs and the rational-choice school, that most influences the framework employed by this study. It too is more concerned with analysing and predicting the behaviour of parties than with explaining their genesis, but its focus is on parties' relations with voters and, crucially, with other parties.

Downs developed a theory of spatial competition to explain parties' behaviour in economic terms. His model envisages politics as comprising a single spectrum, from left to right. Voters have views that are located at some point along this spectrum, and which will be the product of

---

[34] Gaffney (1996), p.1.

[35] Alan Ware, *Political Parties and Party Systems* (Oxford, Oxford University Press, 1996), p.8ff.

their economic interests and general place in society. Voters are assumed, first, to form these views independently of political parties, and, second, to pursue them rationally and self-interestedly. Politicians are assumed to be motivated solely by attaining power. Parties, therefore, are vehicles for the attainment of office; the attraction of votes and the winning of elections is simply a means to that end. Parties thus seek to maximise their vote by casting their electoral net as widely as possible. For Downs, ideology was but an instrument, a mechanism for making parties' policy platforms cohere, and thus allowing voters to cut their "information costs" in selecting the policy options most favourable to themselves. His model assumes that "Parties formulate policies in order to win elections, rather than win elections in order to formulate policies."[36]

A related view of parties' motivation is that of policy-pursuit, a notion addressed by Strom.[37] His conception retains the idea of party leaders seeking votes instrumentally in order to enjoy the private benefits of holding office, but it appreciates that this incentive must be balanced by upholding the wishes of the party's membership and activists, whose goals are likely to be the implementation of particular policies. Party leaders need a committed (and often voluntary) rank-and-file for various reasons, including administration, recruitment, campaigning and, in some countries at least, finance. They cannot afford to disappoint it too often by appearing to put their own office-seeking before that of having party policy implemented. Furthermore, this conflict gives rise to another trade-off that party leaders have to face: between *immediate* vote-seeking, and the *longer-term* attainment of office and/or policy-pursuit. Participation in government is not always the best way to attain a party's policy goals. Particularly if the party concerned is relatively small and likely to be a junior partner in a potential coalition, the possibility of their impact on the government's policy being outweighed by their estrangement from their more ideologically motivated activists may dissuade the party's leaders from taking up office even when they have the chance of doing so. Activists may consider the party's ideology to be better preserved in opposition rather than in government, and to avoid alienating them, the leadership may concur. There is also the danger of association in the eyes of the electorate with unpopular

---

36 Anthony Downs, *An Economic Theory of Democracy* (New York, Harper and Row, 1957), p.28.
37 Kaare Strom, "A Behavioural Theory of Competitive Political Parties", *American Journal of Political Science* vol. 34, no. 2, 1990, pp.565-98.

or painful measures that the government feels impelled to take. These considerations might, therefore, militate towards a longer-term strategy. A party might decide that its policies would be better promoted by eschewing immediate office.

Strom thus proposes three models of party: those of parties as vote-seekers, as office-seekers and as policy-seekers. All are plausible, he suggests, but all have obvious shortcomings as well. What is needed, therefore, is a theory that combines an understanding of all three types of motive and, just as importantly, predicts the circumstances in which one or other will predominate. Thus, Strom's framework assumes party leaders' goals to be primarily office-seeking; they are "political entrepreneurs". Others in the party will also be attracted to the prospect of personal, private benefits obtainable through its patronage. But the inherent scarcity of these benefits will mean that more purposive, ideological goals—that is, the promise of implementing particular policies—will be needed to maintain the active support of most activists and members. So while party elites will usually be willing to make policy compromises to attain office and, instrumentally, to win votes, pressure from the more ideologically motivated rank-and-file will constrain their scope for compromising the party's stated policy goals.

This in turn contains echoes of other important contributions to political science. Tsebelis's notion of "nested games" involves an attempt to explain apparently irrational behaviour. A political actor, he suggested, might be involved in different but simultaneous interaction with other actors, and the effects of juggling different objectives in different arenas could, if the observer was not aware of all the games being played, lead to what appears to be irrational behaviour.[38] On the theory of party behaviour, the understanding that party leaders have different goals, and that these can sometimes come into conflict, is presented most clearly by Harmel and Janda. They envisage a party's leadership as having different goals: *votes*, in competition with other parties; *unity* between itself, its membership and its associated organisations; and *policy* implementation,

---

[38] George Tsebelis, *Nested Games: Rational Choice in Comparative Politics* (Los Angeles, University of California Press, 1990), p.18. For an interesting application of the framework to the differing rates at which four West European social democratic parties had adopted moderate, vote-winning electoral platforms, see Thomas A. Koeble, "Recasting Social Democracy in Europe: A Nested Games Explanation of Strategic Adjustment in Political Parties", *Politics & Society* vol. 20, no. 1, 1992, pp.51-70.

which has at different times forced Social Democratic leaders to bid for support from other parties in order to build parliamentary majorities.

Ideology

The rational-choice approach to explaining party behaviour, within which all these scholars can roughly be located, tends, naturally enough, to downgrade the role of ideas within it. One significant and wider-ranging analysis of social democratic parties' behaviour, which allocates explicitly a role for ideology in an explanatory framework, is Kitschelt's much-discussed *The Transformation of European Social Democracy*. His starting point is a critique of class-based theories of social democratic fortunes, which suggests that such parties are perennially caught between seeking to expand their appeal beyond a shrinking voter base in the working class, and losing their core support in the process.[39] He suggests, however, that the evidence for this thesis is patchy. Instead, he focuses on other variables. One is the party system, envisaged in Downsian spatial terms. Given the existence of social cleavages in the voting public, parties should be able to identify and implement a "rational" positioning on the political spectrum that either maximises their electoral support, captures the "pivot" of the party system (that is, it takes enough votes to preclude the formation of opposing coalitions) or furthers a strategy of "oligopolistic competition" (attacking another left-wing party in order to gain unchallenged access to its voter base). Why some manage to pursue successful strategies and why others choose less successful, "irrational" strategies, Kitschelt attributes to two factors: organisation; and/or ideology, what he calls a party's "discursive tradition".

"The *continuity of ideas* over time", he suggests, "is a critical factor influencing the ways parties stake out electoral appeals and hammer out intraparty consensus."[40] In Spain, for example, the Socialists lacked the long experience of government that most of their sister parties had. Without this, and indeed with most of their leaders and ideologues having been underground or in exile during the long years of Franco's dictatorship, there had been little practical need to moderate the party's radical socialist

---

[39] See, for example, Adam Przeworski and John Sprague, *Paper Stones: A History of Electoral Socialism* (Chicago, University of Chicago Press, 1986).

[40] Herbert Kitschelt, *The Transformation of European Social Democracy* (Cambridge, Cambridge University Press, 1994), p.265. Emphasis in original.

programme. Thus, when faced with left-libertarian and free-market cleavages, its debate was between its old socialists and those who preferred a programme that responded to the new political environment—a battle that was won impressively by the latter in the form of Felipe González's pragmatic, centrist leadership.

In Britain and Sweden, however, the discursive tradition had deeper and more powerful roots. In both cases, Kitschelt argues, the ideology adopted by its main left-of-centre party was grounded in the debate between socialism and liberalism at the end of the 19th century and the beginning of the 20th. This debate was mainly about the size and power of the state and its capacity for redistribution of resources. It created a range of arguments and viewpoints that were acceptable within a party, and it made both SAP and Labour ill-equipped discursively to respond to the non-materialist, non-statist agenda of both the free-market right and the libertarian left. In Sweden during the 1970s and 1980s, the left-libertarian tide was caught not by SAP but by the Communist Left, Green and Centre parties, which all exploited a general disillusionment with the elitism, corporatism and statism of the political system.

This was not the case elsewhere, however. As well as having avoided the necessity of watering down its socialist ideology due to the practical requirements of holding office, the Spanish Socialists could also look further back to their own ideological foundation. Unlike its sister parties in the north, the syndicalist tradition in Southern Europe provided the Spanish party with a collective memory of a non-statist, communitarian political programme to which its modernising forces could refer as they attempted to respond to the free-market and left-libertarian challenges of the 1970s and 1980s. Something comparable was provided by the Christian democratic influence felt by parties in Germany, Austria and the Low Countries. Of course, social democratic parties in these countries may have been hindered by other factors—primarily, internal party organisation and the constellation of forces in the national party system. But, Kitschelt claims, compared to their counterparts in Britain and Scandinavia, there were fewer ideological and discursive obstacles to their modernising and embracing an updated political agenda.

For all its superficial plausibility, there are various problems with Kitschelt's conceptualisation of the relevance of ideology to modern West European social democratic parties. One is that it accommodates so many independent variables, and nods to so many branches of the comparative analysis of party behaviour, that most outcomes could be fitted within it. Thus, as one critic points out, "if parties can lose and gain strategic flexibility so readily, does his [Kitschelt's] theory have any predictive power?"[41] Yet, paradoxically, even the few years that have elapsed since its publication have not been kind to it, in terms of the type of political developments it deals with: witness the current renaissance of the left throughout Western Europe. Indeed, it is questionable whether any single theoretical model can be applicable to so many diverse countries as those Kitschelt attempts to analyse—and even they, as he admits, exclude numerous very relevant cases. However, such an ambitious project was never going to be uncontroversial in its arguments, nor completely accurate in its predictions. Moreover, for the purposes of this study, it has at least two characteristics that make it useful.

First, it combines an appreciation of ideology's salience with estimation of other, less abstract variables, ones that are in large part compatible with the rational-choice-based framework employed in other parts of this examination of the Swedish Social Democrats' European policy. Kitschelt dissociates himself from the rational-choice school of political science, his main complaint being that too many of its assumptions are unrealistic. But some of the tools he uses to measure aspects of internal party organisation and inter-party competition are very similar to, or sometimes identical with, those used by scholars who are bracketed with that approach, such as Panebianco and, especially, Strom. Indeed, Kitschelt uses the standard rational-choice method of taking a definition of rational behaviour as a "baseline", and then using that to gauge the importance of exogenous variables to a given problem.[42] In this sense, his theory provides a coherent "bridge" between different approaches to studying party behaviour, which is helpful to the objectives of this investigation. Indeed, if the evidence we have can be seen as supporting Kitschelt's conception of the importance of ideology, it could add weight to his wider theory.

The model's second useful characteristic is that it deals, albeit relatively briefly, with the issue of European integration for a West European social democratic party in the light of his purported change of voters' preferences. He posits that the EU will pose a dilemma for the centre-left. "On the one hand," he explains,

---

[41] Jonas Pontusson, review of Kitschelt, *The Transformation of European Social Democracy* (1994), *Comparative Political Studies* vol. 28, no. 3, 1995, p.472.

[42] Kitschelt, *The Transformation of European Social Democracy* (1994), p. 256.

left-libertarians have a cosmopolitan orientation that induces them to embrace supranational collaboration. On the other, they fight technocratic governance structures, insist on localised democratic participation, and reject the inegalitarian consequences of open market competition. These preferences lead them to be sceptical of European integration. While this issue may divide social democrats, they cannot turn a vice into a virtue and run on the European agenda as the centrepiece of their voter appeal.[43]

This seems plausible, a priori, when applied to the Swedish case, and indeed to those of its Nordic sister parties. It may be that opposition to EU membership in SAP reflected a left-libertarian rejection of the centralisation of political power that the Union represented. The objective of the next two chapters is thus to test (a) the ideological profile of the SAP's membership and (b) its correlation, if any, with the distribution of attitudes towards European integration within the party. Although, according to Kitschelt, SAP has not been very successful in mobilising the left-libertarian cleavage, some of its internal conflicts in recent years suggest that some sort of non-materialist constituency has a presence within the party. If this is the case, we should be able to see how important it was in explaining its deep disagreement over the desirability of joining the EU.

Alternatively, or perhaps in addition, it may be that, Kitschelt's hypothesis notwithstanding, there is something else in the ideological tradition of Swedish Social Democracy that sits ill with the character of the European Union. Certainly, there are authors who claim to have identified something special, even unique, in its intellectual approach to government. Again, if this is the case, our survey data should be able to identify its influence in contributing to the relative Euroscepticism to be found among the grass-roots of the Swedish Social Democratic Party.

---

[43] Kitschelt, *The Transformation of European Social Democracy* (1994), p.297.

3 The "Swedish model" and Europe: Social Democratic strategy and ideology

Why did Sweden stand apart from the EC for so long? There is one obvious answer: the cold war. But there were other factors as well. Swedish non-participation in supranational integration was not, in the main, forced on a reluctant country; only a minority of Swedes hankered after the opportunity to take their place in the Community alongside their European neighbours. If superpower confrontation did raise a very large obstacle to Swedish EC membership, most Swedes, and Social Democrats more than most, saw no great cause to regret that particular consequence. This reflected a peculiarly Swedish type of nationalism. It was not aggressive in the way that European nationalisms have usually been perceived during the 20th century. Rather it combined various elements—older ones, such as folk myth, and newer ones, including active neutrality on the international stage and immense pride in a famously egalitarian society—that gave Swedish "welfare nationalism" a distinctly Social Democratic flavour. This fusion of Social Democracy with Swedish nationalism was actively nurtured by the party, and it had unavoidably Eurosceptical ramifications.

This chapter presents two faces of SAP, the domestic and the external. A description of the domestic face includes estimation of the organisational strength of the party, and the great political success that was the result. It also examines the most important part of the Social Democratic economic strategy in office, what formed the core of the much-discussed "Swedish model". Discussion of the external face recounts the pursuit of two objectives, free trade and neutrality, that stamped the external policy of Social Democratic governments from 1945 until the 1990s. The relevance of such policies for Sweden's position vis-à-vis the EC is obvious. However, foreign policy dovetailed with internal economic policy, at least according to the sophisticated economic strategy formulated by the Swedish labour movement in the 1950s, and the role of the Social Democratic theory of political economy in shaping attitudes to European integration is assessed. The chapter concludes with a review of how this relationship was reflected in the party's view of the developing EC.

The domestic face of Social Democracy

It is hard to overstate how, in the context of a democratic, pluralist and capitalist society, the labour movement has dominated Sweden—SAP in the political sphere, the trade unions in the economic. The link between the party and organised labour, or at least the blue-collar section of it, has been symbiotic and crucial to the success of each. There is reason to question whether that link is now so vital; conflict between the Social Democrats in office and trade unions broke out more frequently after the 1970s. Nevertheless, the alliance endures. There remains an almost congenital identification of perceived interests between SAP and LO, felt from the highest echelons to the grass-roots of each organisation, even though LO union members' automatic collective membership of the party was ended from 1991. For instance, LO's chair can expect to be elected a full member of the Social Democratic Executive Committee, despite there being nothing in either organisation's constitution that requires it. At the other end of the scale, local trade-union clubs—the level to which the power of collective affiliation to the party was decentralised in the late 1980s—often constitute the Social Democratic presence in smaller towns and, importantly, in the workplace. Some description of the structure and history of this relationship, and the political success it brought for SAP, is necessary if we are to understand its behaviour over the European question.

Internal strength, political success

SAP's founding conference took place in 1889 in Stockholm. The meeting comprised 50 delegates from 14 parts of the country, representing 69 trade unions and political and social organisations. It was decided that the party conference was to be the highest decision-making body; with an administrative committee's brief being only to implement conference decisions and to represent the formative party externally between conferences. The position of party leader was not created until 1907. Quite soon, however, changes were made that strengthened the central leadership. In 1894 the party conference voted to institute a national executive, in which the

Stockholm members acted as an executive committee. In 1900 the national executive was expanded, the post of party secretary was instituted and local organisation was reallocated to around 80 Social Democratic branches, or "labour communes", which were to be "directly subordinate to the party executive, the sole central leadership". While many of these labour communes had been established autonomously by local activists, after the turn of the century their rapid proliferation was largely the work of centrally appointed agitators.[1]

Though its organisation does not obviously favour the grass-roots over the leadership as much as some of its founders perhaps intended, SAP remains—formally, at least—a decentralised party. The party's congress remains constitutionally its highest organ. Congress comprises 350 delegates elected by the regional party organisations, plus the National Executive, a tenth of the party's MPs and certain other representatives.[2] Congress has the power to elect the party leader, the 35 members of the National Executive (plus 15 deputy members[3]), the seven members of the Executive Committee (plus seven deputy members)[4] and certain other bodies, as well as to decide on changes to the party's rules.[5] Patronage plays a relatively small part in the leadership's armoury. Election candidates are nominated by individual members, and regional congresses make the final selection.

However, the leadership is more powerful than it might appear, "with the channels available to the member to influence party policies be-

---

[1] Gullan Gidlund, "From Popular Movement to Political Party: Development of the Social Democratic Labor Party Organization", in Klaus Misgeld, Karl Molin and Klas Åmark (eds), *Creating Social Democracy: A Century of the Social Democratic Labor Party in Sweden* (Pennsylvania, Pennsylvania Universiy Press, 1992), pp.101-3.

[2] These include the party auditors; certain party functionaries; since 1968, the chairs of the party's regional organisations; representatives from Social Democratic Youth, the Federation of Social Democratic Women and the Association of Christian Social Democrats; and, since 1984, one from each of the Swedish Confederation of County Councils and the Swedish Confederation of Municipalties.

[3] These deputy members can attend meetings of the National Executive, but can only vote in the absence of a full member. If this is the case, a pecking order of deputy members applies, so that, say, two full members' non-attendance will permit the top two deputies on the list to vote.

[4] These are all drawn from the National Executive.

[5] Jon Pierre and Anders Widfeldt, "Party Organizations in Sweden: Colossuses with Feet of Clay or Flexible Pillars of Government?", in Richard S. Katz and Peter Mair, *How Parties Organize: Change and Adaptation in Party Organizations in Western Democracies* (London, Sage, 1994), p.815.

ing significantly more limited than is suggested by the party statutes".[6] For example, congress has no power over party finances, and public subsidies to parties has made SAP, like others in Sweden, less dependent on maintaining the active participation of its members, and has thus enhanced elites' scope for shaping policy more independently of grass-roots' views. Although debates at congress are often lively, defeat for motions proposed by the leadership is rare, and the leadership has grown adept at engineering an imprecise outcome to an issue if it senses the possibility of defeat.[7] In any case, the leadership has always had scope for deciding which resolutions to act upon, and in what way. The growth of media attention has also changed the role of congress, making it less a less a mechanism for democratic steering of the party by its membership and more a means of conveying the party's message to the wider electorate. Pressure to avoid embarrassing challenges to the leadership has grown.

This increasingly direct relationship between party and voter has also sharpened the situation of the parliamentary group, which needs to balance its relations with the party organisation (and thus, indirectly, the membership) on one hand, and with parliament (and thus, indirectly, the voter) on the other. SAP's parliamentary party is formally responsible to congress. But the group retains its own budget, and its financial autonomy from the main party is considered strong. It has, however, traditionally been disciplined in support of the party leadership. True, there are signs that Social Democratic MPs have become more independent-minded in recent years, a trend that can be traced back to their first experience of opposition for four decades, in 1976-82, and thereafter, when minority governments became the rule.[8] Nevertheless, open division within the parliamentary group remains rare.

SAP has a considerable grass-roots organisation. The system of automatic collective membership of the party through belonging to an affiliated trade union greatly inflated the nominal total of party members. It exceeded 1 million between 1971 and 1989, and in 1983 reached 1.23 mil-

---

[6] Pierre and Widfeldt (1992), p.342.

[7] Pierre and Widfeldt (1992), pp.342-43.

[8] Anders Sannerstedt and Mats Sjölin, "Sweden: Changing Party Relations in a More Active Parliament", in Eric Damgaard (ed.), *Parliamentary Change in the Nordic Countries* (Oslo, Scandinavian University Press, 1992), p.147. Kitschelt also infers that members of the groups have become less regimented in recent years. Herbert Kitschelt, "Austrian and Swedish Social Democrats in Crisis: Party Strategy and Organization in Corporatist Regimes", *Comparative Political Studies* vol. 27, no. 1, 1994, p.31.

lion—equivalent to around 15 per cent of the entire Swedish population.[9] Until the end of the 1980s SAP's membership as a proportion of the national electorate, at 21.2 per cent, was bettered only by the Austrian Socialist Party's 21.8 per cent.[10] This amounted to around 42 per cent of its voters.[11] In addition, other types of affiliated membership—mainly, members of Social Democratic Youth (SSU)—stood at 44,219 in 1989, having been as high as 73,649 in 1980.[12] Since the decision to end trade-unionists' collective membership was realised by 1992, SAP's membership has numbered around 260,000.[13]

So-called "s-representatives", who number nearly 100,000, disseminate the party message in the workplace. Other types of workplace organisation are the Social Democratic union clubs, containing LO members, and the Social Democratic workplace associations, containing both blue- and white-collar trade unionists, of which there were around 450 by the mid-1980s.[14] At the same time, some observers believe that "the amateur officials of the past have been replaced by professionals. The bureaucratisation of the movement has become pronounced."[15] The present author also noted an impression of professionalisation from several interviewees, especially in northern Sweden. One trade-union activist related how he had noticed changes in the local party when he returned to it after a spell in the 1970s during which his membership had lapsed. For example, whereas before, local-council budgets had been discussed by all the local party, he found on rejoining SAP that they were discussed in the council's committees and then only presented to the full local party for approval.

For him, resentment at the power wielded by what were seen as out-of-touch career politicians and bureaucrats was tied up with local resentment at perceived diktat from the "08s" in Stockholm (a nickname derived from the capital's telephone code). The political marketing consultants imported from Bill Clinton's presidential campaign for the 1994

---

9 Anders Widfeldt, *Linking Parties with People? Party Membership in Sweden 1960-1994* (Gothenburg, Department of Political Science, Göteborg University, 1997), p.91.

10 Mair (1991), p.5.

11 Eric S. Einhorn and John Logue, "Continuity and Change in the Scandinavian Party Systems", in Steven B. Wolinetz (ed.), *Parties and Party Systems in Liberal Democracies* (London, Routledge, 1988), p.179.

12 Pierre and Widfeldt (1992), pp.792-93.

13 Widfeldt (1997), p.91.

14 Gidlund, "From Popular Movement to Political Party" (1992), p.119.

15 Einhorn and Logue (1988), p.181.

Swedish election could be seen as a further example of professionalisation,[16] and it was even more visible in the 1998 election campaign. A huge telemarketing campaign employed 5,000 party workers in 26 regions, each thoroughly briefed with the help of four specially designed manuals. Around half a million voters were telephoned directly (which was only half of what the party had aimed for); 100,000 of those received follow-up calls and direct mailshots. A party official suggested that the leadership's long-term goal was to replace entirely traditional campaign methods, such as door-to-door canvassing, with these modern techniques.

Yet whatever the contemporary problems of the Social Democrats' ties to LO, they have surely contributed to the party's remarkable political success (see table 3.1).

One further constraint on the leadership's freedom of action has become a topic of some political controversy in recent times. The party's close relationship with LO has indubitably been a source of immense strength for SAP, both organisationally and (see the following section) intellectually. However, the modern trade-unions' influence on the party is often said by opposition parties to be excessive.[17] Around SKr20 billion is donated by LO to the party every year towards an election fund—perhaps a fifth of SAP's national income. Perhaps just as importantly, 4,000 LO functionaries are detailed to assisting the party's election campaign, while many of the 230,000 trade-union officials, whose duties employers are legally obliged to allow paid time for, also campaign for the Social Democratic cause. Not surprisingly, LO expects something in return. In September 1996, for example, when a Social Democratic government proposed decentralisation of labour rules and cutting time limits on receipt of unemployment benefit, a leading LO official threatened to withhold his confederation's annual contribution to the party. "The members' money will not go towards financing right-wing policies," its executive announced.[18] Later that month the government significantly watered down its proposal.

---

[16] Nicholas Aylott, "Back to the Future: The 1994 Swedish Election", *Party Politics* vol. 1, no. 3, 1995, p.423.

[17] See, for example, an article by leading figures in the Moderate, Liberal and Christian Democratic parties, Gunnar Hökmark, Torbjörn Pettersson and Sven-Gunnar Persson, "S måste avstå från LO-miljonerna", *Dagens Nyheter* August 9th 1998; also Anders Johnson, *Vi står på vägen* (Stockholm, Timbro, 1998).

[18] *Svenska Dagbladet* September 7th 1996.

Table 3.1 Swedish parliamentary elections under universal suffrage[a]

	v	s	mp	c	fp	kd	m	nyd
1921	4.6	39.4	-	11.1	19.1	-	25.8	
1923								
1924	5.1	41.1	-	10.8	16.9	-	26.1	
1926						b		
1928	6.4	37.0	-	11.2	15.9	-	29.4	
1930[b]								
1932	8.3	41.7	-	14.1	11.7	-	23.5	
1936								
1936	7.7	45.9	-	14.3	12.9	-	17.6	
1939								
1940	4.2	53.8	-	12.0	12.0	-	18.0	
1944	10.5	46.7	-	13.6	12.9	-	15.9	
1945								
1948	6.3	46.1	-	12.4	22.8	-	12.3	
1951								
1952	4.3	46.1	-	10.7	24.4	-	14.4	
1956	5.0	44.6	-	9.4	23.8	-	17.1	
1957								
1958	3.4	46.2	-	12.7	18.2	-	19.5	
1960	4.5	47.8	-	13.6	17.5	-	16.5	
1964	5.2	47.3	-	13.2	17.0	1.8	13.7	
1968	3.0	50.1	-	15.7	14.3	1.5	12.9	
1970	4.8	45.3	-	19.9	16.2	1.8	11.2	
1973	5.3	43.6	-	25.1	9.4	1.8	14.3	
1976	4.8	42.7	-	24.1	11.1	1.4	15.6	
1978								
1979	5.6	43.2	-	18.1	10.6	1.4	20.3	
1981								
1982	5.6	45.6	1.7	15.5	5.9	1.9	23.3	
1985	5.4	44.7	1.6	12.4[c]	14.2	-	21.3	
1988	5.8	43.2	5.5	11.3	12.2	2.9	18.3	
1991	4.5	37.6	3.4	8.5	9.1	7.1	21.9	6.7
1994	6.2	45.3	5.0	7.7	7.2	4.1	22.4	1.2
1998	12.0	36.4	4.5	5.1	4.7	11.9	22.9	

Table 3.1 continued

Notes: Shaded sections indicate party in government from that year or including the whole of it. Years without figures indicate change of government without an election.

a = To second chamber until 1968; to unicameral parliament from 1970.

b = Government of or including Free-Thinkers.

c = Involved joint list with kd.

Key: v = Left Party; s = Social Democratic Party; mp = Greens; c = Centre Party; fp = Liberals; kd = Christian Democrats; m = Moderates; nyd = New Democracy; all names include previous party names.

Source: Misgeld, Molin and Åmark (1992).

Social Democratic political economy

The "Swedish model" was a term "used mostly outside Sweden",[19] and even then it was "never subject to precise definition".[20] However, what perhaps became regarded as the model's most distinctively Swedish characteristic was its welfare state—that is, the reach of universal public services and the extent of resource transfers. Elsewhere in Europe, pensions, sickness insurance, unemployment insurance, child care and even health and education were left to non-state actors such as churches and trade unions. In Social Democratic Sweden, they became the state's direct concern. Public schemes in these fields were implemented largely after the 1951 election and the formation of the Social Democrats' coalition with the Agrarians. Social insurance was designed not as a safety net for those unable to afford private-sector alternatives. Rather, it was income-related, ensuring that the middle classes had as much of a stake in them as the working classes, thus mitigating the former's propensity to revolt against the high tax rates needed to pay for them.[21] The idea, according to Esping-

---

[19] Rudolf Meidner, "The Rise and Fall of the Swedish Model", in Wallace Clement and Rianne Mahon (eds), *Swedish Social Democracy: A Model in Transition* (Toronto, Canadian Scholars' Press, 1994), p.337.

[20] Joseph B. Board, "Sweden: A Model Crisis", Current Sweden No. 410 (Stockholm, Swedish Institute, 1995), p.1.

[21] "The Scandinavian model relied almost entirely on social democracy's capacity to incorporate [the middle classes] into a new kind of welfare state: one that provided benefits tailored to the tastes and expectations of the middle classes, but nonetheless retained universalism of rights. Indeed, by expanding social services and public employment, the welfare state participated directly in manufacturing a middle class instrumentally devoted to social democracy." Gøsta Esping-Andersen, *The Three Worlds of Welfare Capitalism* (Cambridge, Polity Press, 1990), p.31.

Andersen, was to "decommodify" labour: by making these services a citizen's right, rather than a privilege, each would have the opportunity of "exit" from having her position in life determined by her utility in the labour market.[22]

Of course, such wide-ranging public services and transfers were only possible because economic resources were available to fund them. For this reason, and because the functions of the welfare state are not areas into which, even now, the EU has extended its activity (with the partial exception of regional policy), the aspect of the Swedish model that we shall examine in this section concerns economic policy. Because of SAP's near-monopoly of government office in the four decades after the second world war, this part of the "model" was almost synonymous with Social Democratic political economy.

In both theory and application, the labour movement's approach to the productive economy reached a remarkable level of sophistication. Economic strategy was tremendously comprehensive, encompassing detailed analyses of microeconomic and macroeconomic problems and proposing varied but coherent solutions to them. Even an economist like the American, Mancur Olson, who was scarcely associated with the left of the political spectrum, could suggest the quality and quantity of Swedish economists, and the esteem in which they were held in the country, as factors in the country's prosperity.[23] It also pre-empted some characteristics of contemporary "modernised" European social democracy, particularly in its acceptance, even its embrace, of free trade, macroeconomic stability, the limitations of demand-management, the importance of the supply side, and the market's role in wealth creation and resource allocation. In practice, however, such characteristics can be exaggerated. Moreover, some of the model's least "modernised" aspects grew in significance over time and increasingly came into conflict with goals attached to the external face of Social Democracy—above all, the need for access to the EC market.

In the 40 years that followed the second world war, the Social Democrats were unambiguously in favour of maximising economic growth as the basis of improving national welfare, principally through the *"leit-*

---

22 Gøsta Esping-Andersen, *Politics Against Markets: The Social Democratic Road to Power* (Princeton, Princeton University Press, 1985), pp.33-34; Esping-Andersen (1990), pp.21-22, 37-47.

23 Mancur Olson, *How Bright are the Northern Lights? Some Questions About Sweden* (Lund, Lund University Press, 1990), pp.73-74.

motiv" of full employment.[24] At first the party, whose economic policy was strongly influenced by LO's economists, seemed tempted by the type of state planning that was in vogue elsewhere in Western Europe. A government commission investigated the scope for nationalisation, and inflation was controlled through price and credit controls; monetary policy was loose to encourage domestic demand. This strategy was soon supplanted, however, by an alternative that in effect reversed the ways in which the main tools of policy were used. Stabilisation policy—that is, keeping inflation low—was pursued not through microeconomic controls, but through a restrictive macroeconomic framework. Full employment, meanwhile, was not left to macroeconomic demand-management, but to microeconomic measures in the labour market. This was what really made the Swedish model distinctive, at least to non-Scandinavian observers. There were two parts to the microeconomic strategy, and each was the primary responsibility of one wing of the labour movement.

Wage formation, at least in the model's "golden age" during the 1950s and 1960s, was left to the trade unions; arguably the party's most important contribution was to leave bargaining to LO and to keep the state out. The unions' guiding principle, adopted in 1951, was "wage solidarity": to secure for its members equal pay for equal work, regardless of what sector they worked in or the profitability of their employer. There was a normative egalitarian objective to wage solidarity, of course, but there were also harder-headed economic motives. Unionisation of the workforce was and remains very high in Sweden, at 85-90 per cent, and until the 1970s the large majority of unionised workers belonged to LO's member unions. LO's power was rooted in its near monopoly control of the labour supply, and one possible threat to that power was wage drift. If wages came to be set too much by market forces, outside the parameters of the national framework deals for which LO was responsible, the confederation would naturally be marginalised. Another threat, however, was unemployment. If Swedish industry became uncompetitive in the international market, job losses would be the consequence, and this too would create a pool of labour outside LO's control.

The answer, according to the classic LO interpretation, was the "EFO model" of wage formation. Named after three economists who proposed it, the model envisaged a "corridor" within which wage costs could be kept at internationally competitive levels. This "dynamic equilibrium"

---

[24] Assar Lindbeck, *Swedish Economic Policy* (London, Macmillan, 1975), p.228.

was to be found by taking into account both productivity growth in the parts of the Swedish economy that were subject to international competition (which employed around a third of the workforce) and relative prices in the relevant international markets.[25] With those variables considered, wages could be set across the entire economy. Wage drift, and thus cost inflation, were to be limited by the restrictive macroeconomic policy. This would restrain aggregate demand in the labour market by pressing down on firms' profits, which would reduce their capacity for bidding up wages. Wage formation was to be steered mainly through a highly centralised relationship between LO and the Employers' Confederation (SAF), established in the Basic Agreement signed by the two parties at Saltsjöbaden in 1938.

Market-led price signals were thus excluded from allocating the Swedish economy's most important resource, labour. What, then, should replace it, if not state planning? The solution was the party's contribution to microeconomic management of the labour market, and the institution set up by a Social Democratic government to administer this policy was the Labour Market Board. Its remit was to encourage workers to join the most productive parts of the economy. Generous subsidies for retraining and relocation—with some penalties for non-co-operation, too—were the main means through which such economic restructuring was to be achieved. Ailing economic sectors were explicitly *not* to be propped up; indeed, putting upward pressure on wages in the lowest-paying sectors was deliberately designed to force these firms out of business, and thus release their labour for growing sectors.[26] The Labour Market Board's management included representatives from the employers as well as from LO, but, in its recruitment, its operation and its whole ethos, it was unquestionably a very Social Democratic organisation. Rothstein identifies its extremely decen-

---

[25] Gösta Edgren, Karl-Olof Faxén and Clas-Erik Ohdner, *Wage Formation and the Economy* (London, George Allen and Unwin, 1973).

[26] An article in the Social Democrats' political journal in 1961 provides a good example of this policy. It pointed to "the Textile Workers' Union, at this summer's LO congress, urging social measures to deal with the decline of the Swedish textile industry under pressure from foreign competition. It was not a demand for artificial protection in the forms of tariffs, etc, but a demand for social investment in re-education and other appropriate measures to bring unused labour into other forms of production." *Tiden*, "Psychologiska förberedelser för EEC", vol. 53, no. 8, 1961, p.449.

tralised and flexible structure as one of its strengths, but "What was lost in formal control was gained in [Social Democratic] ideological cohesion".27

So important was the active-labour market policy to the economic and political interests of LO and SAP that it can be seen as "*the* major ingredient in the famous class compromise between labour and capital in Sweden in the late 1930s".28 It thus both underpinned Saltsjöbaden and predated the wider strategy of political economy that was formulated 20 years later by two LO economists, Gösta Rehn and Rudolf Meidner; indeed, together with the restrictive macroeconomic policy and wage solidarity, it formed the three main prongs of the Rehn–Meidner plan. The Social Democratic leadership became converted to the plan's merits from the mid-1950s. For now, we need only assess, in theoretical terms, its relevance to Social Democratic views of the EC. Did European integration pose any kind of threat to such a comprehensive, but fundamentally national, economic strategy as was contained in the Rehn–Meidner plan?

The external face of Social Democracy

An appreciation of Sweden's changing position in Europe demands some familiarity not only with the political circumstances prevailing before and during the country's application for membership, but also with the historical context in which those circumstances arose. Sweden is a small country, not in terms of territory but, more importantly, in terms of population. With 8.6 million citizens, it is conscious of its usual classification as a small state in the Union. Sweden has not been considered a major power since its defeat at the hands of Peter the Great's Russia in 1718. From this position, two facts of life constrain the range of policy options open to Swedish governments.

The first concerns economic policy. As a capitalist liberal democracy, but with only a comparatively small home market, selling goods abroad is a *sine qua non* of prosperity. Sweden's economy is by necessity export-oriented, and support for international free trade has traditionally been the method through which exports have been facilitated. Indeed, "Extreme liberalism in the field of foreign economy seems...to be one of

---

27 Bo Rothstein, *The Social Democratic State: The Swedish Model and the Bureacratic Problem of Social Reforms* (Pittsburgh. PA, University of Pittsburgh Press, 1996), p.176.
28 Rothstein (1996), pp.92-93.

the most important preconditions for the economic success of small, developed states in the international system."[29] Yet there are risks in expanding trade beyond a national market. "Nationalizations, confiscations, tariff changes, new non-tariff barriers, differing rates of inflation, devaluations or foreign exchange restrictions"—none of which a single national government, let alone a commercial enterprise, can prevent another national government from implementing—"all render reliance upon foreign markets precarious."[30]

The means of combating such risks, and thus maximising the benefits of international market integration, is "functional integration" of public-policy instruments. Countries agree to share the governance of particular economic sectors, and joint regulation can be seen as a regionally provided public good. It is designed to resolve a collective-action problem, in that it disallows free-riding. For example, a country might be tempted to free-ride by having its exports take advantage of a neighbour's open market, while keeping its neighbours' goods out of its own market; or it might seek to use public money to subsidise its own firms in order to give them a competitive advantage over their international rivals. Common trade and competition rules debar, or at least limit, such free-riding, freeing states from a type of prisoners' dilemma. With market integration facilitated, all countries will benefit from the economic benefits that ensue. But small states will derive most gains: they possess only small home markets, so access to an integrated international one will probably be desirable for their firms.

This, it might be assumed, provides a natural incentive for Sweden to participate, not just in global measures to promote free trade, but also in regional ones. Importantly, however, Mattli argues that the logic of broadening the application of regional public goods does not continue indefinitely. At a certain point, the returns from expanding their application to more policy areas and more countries will begin to diminish. Where this point of diminishing returns falls depends largely on when parts of the area subject to the application of regional public goods begin to feel that the joint governance is too remote to pay sufficient attention to their local

---

29 Hans Vogel, "Small States' Efforts in International Relations: Enlarging the Scope", in Ottmar Höll (ed.), *Small States in Europe and Dependence* (Vienna, Austrian Institute for International Affairs, 1983).

30 Walter Mattli, "Regional Integration and the Enlargement Issue: A Macroanalysis", in Gerald Schneider, Patricia A. Weitsman and Thomas Bernauer (eds), *Towards a New Europe: Stops and Starts in Regional Integration* (Westport, Cn., Praeger, 1996), p.139.

needs.[31] To a certain community, perhaps peripheral to the area to which joint public goods are applied, the attractions of autonomous decision-making may seem to rival or outweigh the advantages derived from eliminating exporters' uninsured risks.[32] Think, for example, of the Icelandic case, in which (thus far) the export advantages to be gained from the country's incorporation into the EU's fisheries regime have failed to offset the perceived costs of being subject to that regime, because such a move would demand that Iceland relinquishes its exclusive control of its own fish stocks.

In Mattli's terms, the Swedish equivalent to the Icelandic fish issue was the second fact of Swedish life: military security—or rather Swedish political elites' perception of what security demanded. In a European state system dominated by big countries, Swedish security policy has had to conform to the conditions created by the relationships between the great powers at any given time. Northern Europe and the Baltic Sea have historically been areas of significant strategic interest for two great European powers, Germany and Russia. Swedish governments have continually sought to avoid entanglement in alliances with either of those powers.

Neutrality

Sweden has long considered itself—and, since 1945 at least, been considered—a neutral country. It has not fought a war since 1814, and there are those who trace the country's policy of neutrality as far back as this. Yet despite this tradition, Swedish neutrality has been comparatively flexible. Swiss neutrality was enshrined in international treaty at Vienna in 1815. Austria's was based on a constitutional law passed in 1955. The terms of Finnish neutrality were determined, albeit implicitly, by the Friendship, Co-operation and Mutual Assistance treaty signed with the Soviet Union in 1948. Swedish neutrality, however, was neither imposed by outside forces nor defined in a treaty or law. Instead, it was a foreign-policy strategy pursued at its governments' discretion, and tailored to the prevailing international situation. The official doctrine was described as "freedom from alli-

---

[31] One of the criticisms of the neo-functionalist school of integration theory was that, implicitly at least, it assumed that such broadening (or "spillover") would continue indefinitely, and disregarded the desire for national autonomy among electorates and governments. See, for example, Stephen George, *Politics and Policy in the European Union*, 3rd ed. (Oxford, Oxford University Press, 1996), p.49.

[32] Mattli (1996), pp.140-41.

ances in peace, aiming at neutrality in war". It was also an armed neutrality: credible national defence was a fundamental part of the policy.

Evidence of the flexibility in Sweden's neutrality can be observed
from the policy's earliest days. Involvement in pre-Napoleonic "leagues of
armed neutrality"—designed, like that set up with Prussia and Denmark in
1800,[33] to safeguard trading rights during European wars—had been ad
hoc arrangements. King Karl XIV Johan, "the first Swedish statesman to
describe neutrality as an enduring ambition",[34] was prompted to do so by
the threat of war between Britain and Russia in 1834; but his policy was
later compromised, as his successors supported Denmark in its confronta-
tions with the German states in 1848-56 and took an anti-Russian stance
during the Crimean war. Before the first world war, the Russian threat became increasingly perceived in Sweden, and elements in the government
were sympathetic towards closer ties with Germany. During the second
world war, while preserving its formal neutrality, Sweden initially gave
significant material support to Finland when it was attacked by the Soviet
Union in November 1939.[35] In 1940-41 it submitted to German pressure
and permitted transits of Hitler's troops to occupied Norway. Neutrality in
this context had a basic, practical objective: to keep Sweden out of the war
"at any price, or rather at almost any price".[36]

Sweden has also intermittently shown interest in alternative security arrangements. The country joined the League of Nations after the first
world war. However, armed neutrality was embraced with renewed enthusiasm from 1936, as international tension grew. At that time, Sweden was
interested in closer Nordic security co-operation, and it revived the idea of
a regional alliance after the war, as tension between the superpowers intensified in 1948-49. But Norway especially felt that such an alliance required
closer ties to the Western democracies than could be squared with Sweden's wish to stand firmly apart from the great powers and their allies. The
following March Norwegian MPs voted overwhelmingly to commit their
country to NATO membership, and Denmark immediately followed suit.

---

[33] Jean Tulard, *Napoleon: The Myth of the Saviour* (London, Methuen and Co., 1984),
p.108.
[34] Krister Wahlbäck, *The Roots of Swedish Neutrality* (Stockholm, Swedish Institute, 1986),
p.11.
[35] Hans Mouritzen, *Finlandization: Towards a General Theory of Adaptive Politics* (Aldershot, Avebury, 1988), p.163.
[36] Nils Andrén, "On the Meaning and Uses of Neutrality", *Cooperation and Conflict* vol. 26,
1991, pp.67-83.

Thereafter the cold war saw Swedish neutrality assume a more rigid form. It was transformed, according to Huldt, from an "elastic" policy into a "permanent" one.[37] Indeed, Sweden became the central pivot in a "Nordic balance" between Finland and the two Scandinavian members of the Atlantic alliance, Norway and Denmark.

Swedish neutrality was a decisive influence in the country's relations with the EC. After the war Sweden joined the Organisation for European Economic Co-operation (OEEC) and received Marshall aid, and in 1949 its government adhered to, without joining, the American-led embargo on high-technology exports to the communist bloc. This was partly to demonstrate, both to other states and its own people, its independence from the Soviet orbit.[38] But at a meeting with the governments of two other neutral states, Switzerland and Austria, in October 1961, Sweden nevertheless agreed that an independent trade-policy, unfettered by the obligations of belonging to a customs union, was indispensable for all three.[39] The reason was clear: the Soviet Union was hostile to the Community, regarding it not simply as an economic association but also a putative political union, and one with intimate ties to NATO.[40]

The Swedish form of neutrality made it particularly sensitive to the Soviet position. Partly because Swedish neutrality lacked a basis in law or treaty, it was imperative that Sweden did nothing that might lead other powers to doubt in any way its commitment to the policy. Non-membership of military alliances was insufficient to fulfil this aim. While the other Scandinavian countries were, in the first few years of the EC's existence, experiencing considerable internal debate over whether they should apply for membership, this was much less the case in Sweden. The Social Democratic prime minister, Tage Erlander, declared in an address to the Swedish Metal-Workers' Union on August 25th 1961, the so-called "metal speech", that Swedish nonalignment

---

[37] Bo Huldt, "Socialdemokratin och säkerhetspolitiken", in Bo Huldt and Klaus Misgeld (eds), *Socialdemokratin och den svenska utrikespolitiken* (Stockholm, Swedish Institute of International Affairs, 1990), p.169.

[38] Mikael af Malmborg, *Den standaktiga national staten. Sverige och den västeuropeiska integrationen, 1945-59* (Lund, Lund University Press, 1996), p.432.

[39] Paul Luif, "Austria's Application for EC Membership: Historical Background, Reasons and Possible Results", in Finn Laursen (ed.), *EFTA and the EC: Implications of 1992* (Maastricht, European Institute of Public Administration, 1990), p.185.

[40] Until 1973 all EC member states were also members of the alliance, and until 1995 Ireland was the only Community member outside NATO.

must be supplemented by a persistent effort to avoid any commitment, even outside the sphere of military policy, which would make it difficult or impossible for Sweden, in the event of a conflict, to choose a neutral course and which would make the world around us no longer confident that Sweden really wanted to choose such a course.[41]

An association agreement was the limit of the government's ambitions, and it submitted an application for one to the Community the following December.

The EC's customs union would dictate the terms of a member's trade, and this was a particular problem for Sweden regarding its relations with the communist bloc.[42] More generally, though, it also violated the principle of "economic defence". The policy's basic aim was to augment the credibility of Swedish neutrality by achieving something close to self-sufficiency in certain products, such as food, arms and textiles. This was designed to persuade other states that Sweden was serious about surviving the disruption to its trade and perhaps the blockade that war in the vicinity might bring. A Commission for Economic Defence had been set up as early as 1928, and it was superseded by similar bodies at various points up to the late 1970s.[43]

Yet Swedish political opinion was not unanimous that Soviet sensitivity excluded the possibility of joining the EC. Hopes, particularly among the Conservative (later the Moderate) and Liberal parties, that neutrality might be compatible with EC membership rose at the beginning and the end of the 1960s, the latter period following the crisis in the EC caused by the insistence of France's President de Gaulle that the national veto remain at the disposal of each member state. In their resolution to the crisis, the Luxembourg compromise of January 1966, the Community's governments agreed that when an issue arose in which a member state considered its vital national interests were at stake, the others would endeavour to find a consensus. In other words, the national veto would remain intact, even when there was no legal basis for it. This encouraged Sweden to think that

---

[41] Cited in Katrina Brodin, Kjell Goldmann and Christian Lange, "The Policy of Neutrality: Official Doctrines of Finland and Sweden", *Cooperation and Conflict* vol. 3, 1968, p.28.

[42] For instance, it was claimed that inclusion in the customs union would prevent Sweden from selling lead piping to the Soviet Union. Daniel Tarschys, "Neutrality and the Common Market: The Soviet View", *Cooperation and Conflict* vol. 11, 1971, p.71.

[43] Ebba Dohlman, *National Welfare and Economic Dependence: The Case of Sweden's Foreign Trade Policy* (Oxford, Clarendon Press, 1989), pp.53-56.

a more traditionally intergovernmental, less supranational type of organisation might be evolving. In July 1967 the Swedish government submitted an "open application"[44] to the Community, in which negotiations began on the terms of Sweden's relations with it. Membership was not made the goal of these negotiations, but nor was it ruled out. In an article in *Tiden*, the Social Democratic political journal, a departmental secretary in the Swedish Ministry of Finance discussed the renewed exploration of membership. Although he acknowledged that de Gaulle's protest had not stopped integration in the EC, he concluded: "In the current situation, in which federalist and supranational ideas are receding, *the influence of the different member states* in co-operation in Brussels has been accentuated."[45]

The article was submitted to *Tiden* on November 15th 1967. A fortnight later came de Gaulle's famous press conference, in which he blocked Britain's application for a second time. The Swedish government decided to wait before responding, and even after the six EC governments accepted the Werner report on monetary union in December 1969, the Social Democratic prime minister, Olof Palme, seemed still interested in some form of membership. During a tour of European capitals, he told a British newspaper in April 1970 that Sweden could accept the economic obligations of membership, and even some political ones, as long as they fell short of co-ordinating defence and security policy.[46] But publication of the Davignon plan for closer co-operation between the EC's member states on foreign policy in October 1970 in effect ended Swedish hopes of membership. With the credibility of Swedish neutrality now at issue, the EC's stated aims were as important as the substance of its contemporary character. As Palme put it in 1971, "Words are a political reality."[47] On March

---

[44] Klaus Misgeld, "Den svenska socialdemokratin och Europa—från slutet av 1920-talet till början av 1970-talet. Attityder och synsätt i centrala uttalande och dokument", in Bo Huldt and Klaus Misgeld (eds), *Socialdemokratin och den svenska utrikespolitiken* (Stockholm, Swedish Institute of International Affairs, 1990), p.207.

[45] Stig Lindblom, "Bollen ligger inte hos oss: EEC och anslutningsformerna", *Tiden* vol. 59, no. 10, 1967, pp.601. Emphasis in original.

[46] Cynthia Kite, *Scandinavia Faces EU: Debates and Decisions on Membership 1961-94* (Umeå, Department of Political Science, Umeå University, 1996), p.141.

[47] Cited in Carl-Einar Stålvant, "Rather a Market than a Home, But Preferably a Home Market: Swedish Policies Facing Changes in Europe", in Finn Laursen (ed.), *EFTA and the EC: Implications of 1992* (Maastricht, European Institute of Public Administration, 1990), p.140.

18th that year Palme informed the Swedish Foreign Affairs Council[48] that the government had decided to rule out the option of membership. Party unity was, by some accounts, a significant consideration for Palme. His finance minister, Gunnar Sträng, was particularly hostile towards the EC,[49] and—as we will see—there would very probably have been a large part of the labour movement that would have agreed with him.

Political and military developments in the 1980s made Swedish membership seem even more inconceivable. The Single European Act, agreed in December 1985, took significant steps towards supranationalism in its introduction of qualified majority voting in the Council of Ministers for matters pertaining to the completion of the EC's internal market.[50] Moreover, the Single Act gave legal status to foreign-policy co-ordination among the member states. Security developments—the stationing of new missiles in Western Europe by NATO, the reinforcement of Soviet and American forces in Northern Europe, and the repeated "encounters" between Swedish naval forces and suspected Soviet submarines[51]—persuaded the Swedish Social Democratic government and most of the opposition that the credibility of the country's neutrality needed to be further emphasised.

Free trade

While neutrality comprised the foundation of Sweden's security policy, the basis of its trade policy was that there was no contradiction between this

---

[48] This body, established in 1921, is chaired by the king, and comprises the speaker of the Riksdag, nine MPs (including the main party leaders) and nine reserve members (also MPs). "The Council can be seen as a platform for the opposition to make itself heard on foreign policy matters, and has sometimes also been used by the government to mark a distinct line." Arne Halvarson, *Sveriges statsskick: fakta och perspektiv*, 10th ed. (Stockholm, Almqvist & Wicksell, 1995).

[49] Jakob Gustavsson, *The Politics of Foreign Policy Change: Explaining the Swedish Reorientation on EC Membership* (Lund, Lund University Press), pp.50-51.

[50] "[The SEA] has not made Swedish membership any easier, since it formally recognises the link between inter-governmental political cooperation and treaty-based integration." Carl-Einar Stålvant and Carl Hamilton, "Sweden", in Helen Wallace (ed.), *The Wider Western Europe: Reshaping the EC-EFTA Relationship* (London, Pinter, 1991), p.211.

[51] Stålvant (1988), p.18. Although one Soviet submarine did run aground in Swedish waters, a government report published in early 1995 suggested that many of the Swedish navy's contacts with supposed Soviet submarines in the 1980s had in fact been with swimming mink.

and support for the freest possible trade in Western Europe.[52] Like Britain and the other Nordic states, free trade in Europe was thus Sweden's preferred alternative to Community membership. Such was the importance of the British market for their exports that a tariff barrier between the Scandinavian countries and Britain—the effect if either they or Britain had joined the EC without the other—was highly undesirable for them. For these reasons, the British "grand design" for a wide European free-trade area (WFTA), encompassing all the member states of the OEEC, received solid support from the Nordic governments when it was first presented in October 1956. But it was rejected by France, which pushed the EC instead.

One alternative for Sweden, a Nordic customs-union, was superseded by establishment of the European Free Trade Association (EFTA). The Stockholm Convention of January 1960, in which the seven member states of the association (Austria, Britain, Denmark, Norway, Portugal, Sweden and Switzerland) agreed to remove tariffs between them on manufactured goods (which was achieved within six years), represented a confluence of all their interests. Its minimal, intergovernmental institutions were acceptable to the neutral states, and to Britain, Norway and Denmark. A free-trade area, without a common external tariff, permitted the member states to maintain their divergent trade regimes with the outside world. Most importantly, however, EFTA represented an attempt by the member states to create a stronger negotiating position with the EC.[53] Certainly, EFTA suited Swedish interests. In the Riksdag, only the Communists voted against the convention.

But even as the EFTA countries agreed in June 1961 to defend their common interests, some of their number sought to defect from the association. Britain first decided to apply for EC membership the very next month. This was taken badly in Sweden. *Tiden* declared dramatically: "The truce [with the Six] that Sweden and the other neutral countries believed they had worked out through the EFTA agreement, has...become a war."[54] Denmark and Norway, meanwhile, the latter feeling compelled by the prospect of two major trading partners decamping to the other side of the

---

[52] This was with the exception of those sectors in which it was thought that self-sufficiency was needed. Indeed, it was argued that the wealth that free trade created was itself necessary to maintain a credible policy of economic defence. Gunnar Sjöstedt, *Sweden's Free Trade Policy: Balancing Economic Growth and Security* (Stockholm, Swedish Institute, 1987), p.34.

[53] Miljan (1977), pp.143-44.

[54] *Tiden*, "Det europeiska alternativet" (1961).

EC's tariff wall, moved with Britain in each of its three applications. And while it remained mindful of its very different security considerations, Sweden was also interested. Like Switzerland and Austria, it inquired about associate membership in 1962; in 1966 its government wrote to the Community, asking that in the forthcoming negotiations Swedish interests be borne in mind;[55] and in July that year there was relative domestic consensus behind the "open application" for membership.

After de Gaulle's second veto in 1967, however, accession to the EC seemed closed as an option for Britain and Scandinavia. Once again, therefore, a Nordic alternative was mooted, this time by Denmark. The proposal was for a Nordic common market, Nordek, and a Committee of High Officials from the respective governments submitted a report on the plan in July 1969. There were important differences between the Nordic countries over the plan's purpose, however; and in any case, the possibility of EC membership for Denmark and Norway came alive again after April 1969, when De Gaulle's resignation prompted a third British application to the Community. Sweden too began talks with the EC in 1970, and seemed to believe that inclusion in the customs union was not necessarily incompatible with neutrality, even if full membership was. Certainly, big business was amongst the strongest advocates of closer relations between Sweden and the EC.

But once developments within the Community (the Werner and Davignon plans) had persuaded the Swedish government to renounce full membership the following year, the alternative of bilateral free-trade agreements between the remaining members of EFTA and the EC was accepted as the best way to advance Sweden's interests.[56] The agreements were duly reached in 1972 and implemented the following year. The Swedish trade minister, Kjell-Olof Feldt, declared: "[T]he issue that during the last 15 years has stood at the forefront of our trade policy has been resolved."[57] In 1984 the final provisions of the agreements came into effect:

---

55 Carl-Einar Stålvant, "The Exit vs Voice Option: Six Cases of Swedish Participation in International Organisations", *Cooperation and Conflict* vol. 11, 1976, p.43.

56 "It was a widespread notion within the Swedish political elite that the country through this arrangement had gained all the economic advantages of...membership without having any of political drawbacks." Magnus Jerneck, "Sweden—The Reluctant European", in Teija Tiilikainen and Ib Damgaard Petersen (eds), *The Nordic Countries and the EC* (Copenhagen, Copenhagen Political Studies Press, 1993).p.25.

57 Quoted in Gullan Gidlund, *Partiernas Europa* (Stockholm, Natur och Kultur, 1992), p.38.

as the last tariff barriers to trade in industrial goods between EFTA and the Community disappeared, it could be argued that a WFTA, the object of British and Nordic policy since the end of the second world war, had been realised. However, once again the EFTA states found themselves immediately having to adapt to new circumstances beyond their control.

In June 1985 the European Commission proposed nearly 300 measures to facilitate free movement of goods, services, labour and capital inside the ring-fence of the Community's common external tariff. What the countries of EFTA feared was their exporters becoming disadvantaged in the single market compared to EC firms. They were not only concerned that tariff- and quota-free trade with the EC was limited to industrial goods; non-tariff barriers were another worry. The single-market programme was to be founded on two broad principles: common regulatory standards in some fields, and mutual recognition of member states' standards in others. If these principles did not apply to the EFTA states, firms based within the EC, and thus attuned to its regulatory standards, might have an advantage in trading within it over those firms based outside, whose home country's standards might differ from the Community's.[58] Anita Gradin, the Swedish trade minister, admitted that "the decisions by the European Community with regard to the Internal Market have created a feeling of urgency among the EFTA countries to find a new basis for a durable relationship with the EC."[59] This concern was aggravated by a perception, in Sweden especially, that the EC might be assuming a more protectionist stance.[60] The Uruguay round of the General Agreement on Tariffs and Trade (GATT) also seemed to be faltering, threatening the advancement of freer global trade.

EFTA's growing trade-dependence on the EC—between 1972 and 1986, trade between them quintupled—only made these worries more acute. Sweden was particularly dependent on trade with the EC (see table 3.2). At Sweden's instigation, then, in April 1984 the two associations agreed "further actions to consolidate and strengthen cooperation, with the aim of creating a dynamic European economic space of benefit to their

---

58 See, for example, Esko Antola, "EFTA-EC Relations After the White Paper", *EFTA Bulletin* vol. 28, no. 3, 1987, p.3.
59 Anita Gradin, "West European Integration: The Swedish View", *European Access* vol. 3, no. 2, 1989, p.22.
60 Carl Hamilton, "Protectionism and European Economic Integration", *EFTA Bulletin* vol. 28, no. 4, 1987, pp.8-9.

Table 3.2 Percentage of Swedish foreign trade

	1960	1970	1990
EFTA	29	42	18
EC	36	31	55
North America	10	8	9
Others	24	19	18

Note: An important factor in Sweden's changing trade profile was the enlargement of the EC and the concomitant shrinking of EFTA.
Source: Statistics Sweden, cited in Gustavsson (1988), p.81.

countries."[61] This marked the beginning of the so-called Luxembourg process, which after 1989 became known as the Delors/Oslo process.

Moreover, Sweden took its own steps. Propositions from the Social Democratic government and contributions from the opposition parties led, in May 1988, to a position that all bar the Communist Left could agree. Although EC membership was not to be an objective at that time, Sweden would seek to avoid its citizens and businesses becoming in any way disadvantaged in the SEM. Moreover, the government set up a new ministerial committee on European affairs, comprising the prime minister, the foreign minister and the ministers of finance, foreign trade and industry; an advisory Council of European Affairs, in which the government, business, trade unions and academics were represented, was set up.[62] All future legislation was to be vetted for its compatibility with EC norms, and more than 20 EC directives were adopted into Swedish law.

This new phase of EC–EFTA negotiations were to prove difficult for the EFTA countries. In December 1989 they agreed that the *acquis communautaire*, the Community's accumulated body of law, would be the basis of common trade relations, but they still demanded "a more structured partnership with common decision-making and administrative institutions in order to provide for the joint shaping and making of future EC rules".[63] The EC demurred, however, and by November 1990 the EFTA countries had bowed to the inevitable and had dropped their demand for

---

[61] *EFTA Bulletin* vol. 25, no. 2, 1984, p.7.
[62] Gustavsson (1988), p.59-61.
[63] *EFTA Bulletin* vol. 29, no.4, 1989 / vol. 30, no. 1, 1990, p.3.

equal decision-making rights, in return for certain assurances.[64] Yet even after agreement was finally reached on what was now referred to as the European Economic Area (EEA), the objections of the EC's Court of Justice and the Swiss electorate had to be overcome. A revised EEA eventually came into force on January 1st 1994, a year behind schedule.

The confluence of external and internal faces: Social Democratic ideology and European integration

The Swedish Social Democrats' response to the Briand plan for a European political union was fairly unequivocal. "The idea of a European union...[and] the French foreign minister's initiative", declared the party newspaper, "can be interpreted as a step onward towards limiting national sovereignty and solidarity between peoples, things for which Social Democrats have always campaigned."[65] This comment was made as long ago as 1937. Does this mean, then, that there is something fundamentally problematic about the notion of European integration for the party's ideology? If so, in what way?

Social Democratic political economy and the EC

Interest in European integration grew in Social Democratic circles as, during the 1960s, the prospect of Swedish associate or even full membership of the EU became increasingly feasible. Eurosceptical argument within the party was expressed most forcefully in a book by Tord Ekström, an LO economist, Roland Pålsson, a commentator on economic and foreign affairs, and Gunnar Myrdal, one of the main architects of Social Democratic political economy and a former Swedish trade minister. Their book, *Vi och Västeuropa* (We and Western Europe), attacked the possibility of Sweden's drawing closer to the EC. One of their main objections concerned economic policy.

---

[64] These included the right to be consulted at an early stage in EC policy-making, transitional arrangements and a safeguard clause, providing for the suspension of any part of the agreement in exceptional circumstances Clive Church, "EFTA and the European Community", European Dossier 21 (London, PNL Press, 1991), p.15.
[65] Editorial in *Socialdemokraten*, May 20th 1930. Cited in Misgeld (1990), p.197.

Ekström, Myrdal and Pålsson suggested that tariff barriers were not actually very important in hindering trade between developed West European countries (although the level of the Six's common external tariff had prompted some concern in the Riksdag when it agreed to the EFTA convention[66]). Rather, it was because price reductions, which removal of tariff barriers should have prompted, were not always passed on to consumers, due to a lack of genuine competition within the common market. In any case, the EC was less a vehicle for promoting free trade as one with powerful protectionist and mercantilist instincts. Indeed, partly because of this, the authors feared the effect that an influx of continental cartels into the Swedish market would have on domestic enterprises and consumers: "In comparison with Swedish practice, the Six have greater restrictions on 'low-priced goods' from under-developed countries, and this brings risks of cartels having freer play in the Community than would otherwise be the case."[67] An editorial in *Tiden* in 1961 took the same line. Discussing possible Swedish membership of the Community, it was clear, it argued,

> that accepting the Treaty of Rome's outer tariff walls would in many respects contradict essential Swedish interests. For a very long time, Sweden has implemented a policy of free trade with the rest of the world. The Six's tariff walls will be markedly higher than our existing Swedish ones and would thus have the effect of forcing Sweden to raise tariffs against other countries.

The third world would be especially unfavourably hit by any such development, violating another Swedish foreign-policy goal.[68]

In fact, the problem of the EC's protectionism diminished in the 1970s as the Community became more open to trade with the outside world. In any case, the accession of Denmark and especially Britain in 1973 meant that two important markets for Swedish goods were now within the customs union; Sweden could probably not afford to be quite so selflessly concerned about the third-world development, which pursuit of global free trade was said to represent. In any case, Ekström, Myrdal and Pålsson made clear that they had more fundamental objections to Swedish EC membership.

---

66 *Nordisk Kontakt* 6, 1960.
67 Ekström, Myrdal and Pålsson (1962), p.121.
68 *Tiden*, "Neutraliteten och De Sex", vol. 53, no. 7, 1961, pp.386.

Party, nation and neutrality

The relationship between economic class and nation has often been a tricky one for the European left; but it has probably been less a problem for SAP than it has for its sister parties elsewhere. Indeed, the fusion of class-based internationalism with a strong sense of national pride has been among the party's most significant political achievements. As early as 1900 the party's first leader, Hjalmar Branting, was stressing that "the country is worth defending"—partly to distinguish the Social Democrats' position from that of the Communists, whose allegiance they suggested was actually to Moscow.[69] According to Tingsten, the party came to believe that "A socialist movement that enjoyed a tradition of successful activity within a state had to be national—if not because of patriotic sentiments then in order to protect socialism."[70] In addition, the party's emphasis on class as the basis of its politics has been much less pronounced than has that of other socialist parties in Europe. Esping-Andersen argues that in Scandinavia, "Social democracy...distinguished itself by the decision to subordinate class purity to the logic of majority politics. The organisation moved from 'working-class politics' to 'people's party'; its platform addressed the 'national interest' rather than the old proletarian cause."[71] Indeed, it is arguable that the cross-class appeal that SAP soon adopted, and the arrogation to themselves of the mantle of the "national party", has also contributed greatly to their self-perception, and thus also to the policies that the party's internal debate has generated.

The idea of "the Swedish people" is, claims Micheletti, "an organic concept stressing the collectivity rather than the individual as the basic unit of society", and the Social Democrats exploited it skilfully, expanding their traditional use of the term to denote working-class solidarity.[72] It was Per Albin Hansson, SAP's leader from 1928, who coined the term *folkhemmet*, the "people's home", to describe his party's vision. Indeed, it was no coincidence that SAP's appropriation of the nationalist im-

---

69 Peter Wallensteen, "Socialdemokratin och säkerhetspolitik: några kommentarer", in Bo Huldt and Klaus Misgeld (eds), *Socialdemokratin och svensk utrikespolitik. Från Branting till Palme* (Stockholm, Utrikespolitiska institutet, 1990), p.183.
70 Herbert Tingsten *The Swedish Social Democrats: Their Ideological Development* (New Jersey, Bedminster Press), p.569.
71 Esping-Andersen (1985), p.8.
72 Michele Micheletti, *Civil Society and State Relations in Sweden* (Aldershot, Avebury, 1995), p.58.

age occurred at the same time as nationalism was establishing itself in an altogether more aggressive form elsewhere in Europe; the process may have been fostered to steal some of the nationalist right's clothes, and so stifle its development in Sweden.[73] Whatever his motivation, Hansson cleverly managed to fuse the symbols of party and nation. On Sweden's national flag day in 1934 he hoisted the national emblem over a metaphorical people's home.

Other Social Democratic thinkers did much the same over the party's years of dominating Swedish politics. Gunnar and Alva Myrdal couched their discussion of the population question, dramatically, in terms of the survival of the Swedish nation. Erlander developed the concept of the people's home into the "strong society", a "mystical merging and identification of state and society", to the extent that he implied that "to oppose the public sector is not merely to object to government activity, but to question the choices of society itself."[74] In the 1985 election, Tilton describes how "the Social Democrats wrapped the welfare state, the Swedish nation, and their party into a single package and sold it to the electorate."[75] Moreover, the Social Democratic brand of Swedish nationalism helped to shape Sweden's behaviour on the international stage, particularly when it came to relations with the EC. Neutrality played a crucial part in Sweden's international role—not least in the way it shaped Social Democrats' own perception of that role.

Ekström, Myrdal and Pålsson argued that while associate membership would be technically compatible with neutrality, in practice it would fundamentally alter the perception of Sweden's security policy abroad, thereby fatally undermining its credibility. This was due to the "obvious fact that the Community involves an economic and political fusion of states that belong to the same military alliance [NATO]".[76] The authors also pointed to what they saw as a group of European countries that had traditionally been neutral between East and West.

Why had neutrality taken on such importance for the Social Democrats? As well as being electorally popular, the policy had its domestic uses. Because "economic neutrality" was deemed to require the potential for self-sufficiency in some sectors, including food, it was a

---

[73] Henrik Bergren, "Rätt diagnos, fel recept", *Dagens Nyheter* May 13th 1997.

[74] Tilton (1991), p.187.

[75] Tilton (1991), p.280.

[76] Ekström, Myrdal and Pålsson (1962), p.136.

means to justify the generous agricultural subsidies that helped smooth the Social Democrats' parliamentary co-operation with the Agrarians and its later incarnation, the Centre Party, as during the coalitions of 1936-39 and 1951-57. Moreover, support from the Swedish communists (who became the Communist Left in 1967, and the Left in 1990) was often enough to secure parliamentary majorities for SAP governments, as in 1960-73 and 1982-88, and the Communists were jealous of any threat to neutrality. Occasionally it could also be used as a vote-winning stick with which to beat the opposition. In 1959, for instance, Krushchev, the Soviet leader, cancelled a visit to Sweden, blaming critical comments by the Conservative Party leader. The Social Democrats exploited this by questioning the Conservative leader's political judgment, and they did much the same in the 1970s and 1980s when his successors complained about alleged Soviet submarine intrusions into Swedish waters.[77]

But there was more to the Social Democratic attachment to neutrality than vote-seeking and parliamentary manoeuvring. The concept also developed into a tool for shaping party unity, as it reflected the advance of a more radical ideological strain within SAP. According to Stråth, this was first reflected in reaction against a post-war campaign—propagated by Herbert Tingsten during his tenure as editor of the country's main broadsheet, Dagens Nyheter—to have Sweden follow Denmark and Norway into NATO. In a series of newspaper articles, leading Social Democratic radicals outlined a "third way" for the country's foreign policy: anti-American, strongly anti-colonial, pro-third-world and, if not overtly pro-communist, content to keep criticism of the Soviet Union "pushed into the background". Neutrality thus "took, in this third-way form, a hegemonic character" in the party.[78] It reached a new stage from the late 1960s, when Palme's more "active" neutrality was directed especially against American involvement in Vietnam.[79]

Stråth argues that this fusing of security policy with political ideology was, in turn, partly because the "third way" captured and articulated even deeper-seated traits of the party's identity.[80] Largely because the re-

---

[77] Andrén (1991), pp.75-76.

[78] Bo Stråth, Folkhemmet mot Europa. Ett historisk perspektiv på 90-talet (Stockholm, Tiden, 1992), p.201.

[79] Andrén (1991), p.80.

[80] Later the third way became still broader as a concept, with some Social Democrats using it to connote a Swedish path between capitalism and socialism. See, for instance, Rudolf Meidner, "Our Concept of the Third Way: Some Remarks on the Socio-Political Tenets of

formist, parliamentary route to power had proved so successful for the party, Social Democrats developed an ideological connection between national sovereignty and democracy. Ekström, Myrdal and Pålsson insisted that ultimately Sweden's choice in European policy was driven by political, not economic considerations, and that the basic political character of the EC was undemocratic. The Community's leading institutions were not subject to proper parliamentary control; nor were the administrative agencies that appeared to be playing a large part in its governance of the common market. Agreements forged between independent, sovereign states in specific, limited fields were, they argued, one form of co-operation. These would permit the contracting parties to be effectively controlled through popularly elected national assemblies. But the EC's system demanded

> that we shall give up our national sovereignty in advance and permit supranational bodies, in the interest of "Europe", to make binding rules for us in Sweden without subjecting them to the test of constitutional procedure, which puts them beyond parliamentary control and, in the final analysis, beyond the control of the electorate.[81]

Indeed, Ekström, Myrdal and Pålsson's belief that Sweden's level of economic, social and political progress compared favourably to that in the Six runs through their book. Offering a polemic against the idea of a unifying European culture, they suggested that Swedish political culture had more in common with Australia's than Southern Europe's. "It is, above all," they argued,

> the firmly Protestant countries that have come furthest in economic and all other development...Evidence for this generalisation can be seen, for example, in how it is largely the Protestant countries that have succeeded in implementing effectively an income tax that most people pay honestly...Everyone knows that democracy is more transparent, entrenched and effective in the Anglo-Saxon emigrant countries, much as it is in Britain and Scandinavia.[82]

Political integration was understandable from the viewpoint of the Six, they argued. But those countries' attempts go beyond the national frame-

the Swedish Labour Movement", *Economic and Industrial Democracy* vol. 1, 1980, pp.343-70.

81 Ekström, Myrdal and Pålsson (1962), p.20.

82 Ekström, Myrdal and Pålsson (1962), pp.34-35.

work in solving the social and political problems they shared was not itself a reason for Sweden's participation. After all, Sweden had had great success in making social and economic progress from the basis of the nation state.[83]

Nor was these authors view an isolated one in SAP. A year before their book's publication, *Tiden* declared that the EC epitomised a "Europe that can be likened...to the backwardness of the French and Italian nations in their economic and social structures".[84] "For 20 years", declared another editorial, "the thought of a united Catholic Europe inspired Konrad Adenauer."[85] Rather, the achievements of the Swedish model would be "lost with [EC] membership, according to dominant opinion in SAP".[86] The clear sense of pride, even of mission, that combined national and Social Democratic identity can be see in various other forms. Inga Thorsson, who became a leading opponent of Swedish membership of the EU in the debate that preceded the referendum in 1994, expressed just such sentiments in 1962. She acknowledged that some other countries in Europe might regard Sweden's lack of enthusiasm about integration as being selfish and narrow-minded. She was unmoved.

> We disapprove of some European states' inability to discard their great-power dreams and abandon the remnants of colonialism in a dignified way. We disapprove of the Catholic countries' view of rational family-planning, because we believe it to be one of the only ways towards an acceptable standard of living for masses of starving human beings...I believe that one must be able to stand up and be counted, even if it is unpopular.[87]

Meanwhile, the idea that the EC could advance socialism, through peace and economic planning, "played no role in the Swedish debate. The arguments were thought to be irrelevant to Sweden."[88]

---

[83] Ekström, Myrdal and Pålsson (1962), pp.43-45ff.

[84] *Tiden*, "Det europeiska alternativet", vol. 53, no. 5, 1961, pp.313-17.

[85] *Tiden*, "Bonn och Europa", vol.3, 1960, pp.181-84.

[86] Misgeld (1990), p.202.

[87] Inga Thorsson, "Sverige, svensk socialdemokrati om omvärlden", *Tiden* vol. 54, no. 5, 1962, pp.262-67.

[88] Misgeld (1990), p.205.

Conclusion

It is not hard to see, then, at least two of the most important influences that have been described in this chapter reflected in the Social Democratic debate on the merits or otherwise of Sweden's joining the formal process of European integration. Clearly, there is wide acceptance, even in the more radical sections of the Swedish left, that the country exists in an international market economy, and that free trade, at least in its basic form, is desirable. It is conceivable that the perception of the EU as a mercantilist trade bloc persuaded some Social Democrats that to join would violate, not promote, their free-trading tradition.

On the other hand, Social Democratic ideology has undoubtedly become infused with national sentiment. This is probably inevitable in a country where a single political movement has been so dominant for so long. It has occurred thanks to the sheer success of the party, both electorally and in its presiding over what for many years was one of Europe's most successful economies, and certainly one of its most developed and comprehensive welfare states. It was also reinforced by the experience of neutrality, of the need for a common national position in facing the outside world, in order to maintain what sometimes seemed like a precarious existence between two opposing power blocs. Lawler, in a recent article exploring the normative aspects of Nordic Euroscepticism, argues that "At the centre of the Scandinavian worldview is a positive model of the internationalist sovereign state",[89] and all these factors—some of which apply throughout the region, all of which apply in Sweden—have contributed to making this the case. It has also been achieved, in some respects, by the conscious design of Social Democratic leaders and ideologues; the sense of political, social, economic and cultural superiority vis-à-vis the EC is, in a sense, mythical. But, as Lawler acknowledges, "myths matter" in a country's foreign policy.[90]

Yet we cannot just accept the suggestion of a positive correlation between, on one hand, support for the nation state and its achievements, and, on the other, a particular attitude to European integration. Few would argue that, say, French Socialists have a negative attitude towards their nation state, yet most are strongly supportive of the EU. Even in the

---

[89] Peter Lawler, "Scandinavian Exceptionalism and European Union", *Journal of Common Market Studies* vol. 35, no. 4, 1997. p.568.
[90] Lawler (1997), p.571.

Swedish context, we cannot draw any automatic inference about what ideological background means for Social Democrats' perspective on the EU. "Welfare nationalism", as Elvander puts it, can work both ways. It can be either "isolationist" in form, and seek to keep Sweden from contamination by the rest of Europe; or it can be "evangelical", and look to export the achievements of Swedish Social Democracy to the rest of the Union.[91]

More survey research is clearly required if the ideological connections between the Swedish Social Democratic tradition and support or opposition to European integration is to be established. This is what the following chapter presents.

---

[91] Nils Elvander, "Självbelåten välfärdsnationalism styr nej-sidan", *Svenska Dagbladet* November 6th 1994.

4 Arguments and evidence: the contours of division

In the previous chapter, we examined the Social Democrats' ideological heritage, and how this might condition the party's approach to European integration. But how important to the party's divisions over EU membership were the distinct ideological elements that we identified? Can existing ideological conflict be correlated with support for or opposition to Sweden's joining the Union? The data contained in this chapter allow us to generate certain hypotheses about which were the most important issue cleavages in SAP's European debate, and to test their salience. For example, was it simply a matter of left against right, or materialists against post-materialists, or nationalists against internationalists?

If it can be seen that the division over whether or not Sweden should join the Union corresponds significantly to another fault line in the party's ideological profile (over free trade or security policy, for instance), then we may conclude—as these fault lines all predate the division over the EU—that SAP's European divide could actually be the expression in new form of an old cleavage within the party. This would go a long way towards an explanation of why the party was so divided over the leadership's decision to change policy towards the EC in 1990. Of course, it would not answer our other research question, about why the leadership felt it had to change policy at that moment and so rapidly. However, if such a conclusion could be made, and a connection established between the split over Swedish accession and a pre-existing division among Social Democrats, then it would also lend strong support towards a more structuralist view of political behaviour. We might infer that Social Democrats were responding less to the EU *per se*, and more to its impact on older patterns of political culture within SAP.

Our first step, therefore, is to outline our hypotheses. These are derived from three separate elements of the empirical research. First, they will be justified with evidence from campaign propaganda and from the interviews conducted in November and December 1993 and November

1994.[1] More systematic is the second element of research, an analysis of the motions on the EU presented to the 1993 party congress. These methods are essentially preliminary. Any quantitative analysis of political behaviour must have some sort of theory to guide it initially, even a tentative one, and these qualitative techniques serve to shape lines of inquiry. Thereafter, the salience of these hypothesised fault lines will be tested more rigorously and systematically in the questionnaire survey.

Ideological cleavages within SAP: six hypotheses

Hypothesis 1: "isolationists" v "missionaries"

The most obvious explanation for a division amongst party activists over questions of supranational integration is that it signified a split between nationalists and internationalists, between, on one hand, people who opposed integration because their primary goals were intrinsically predicated upon the nation state and, on the other, those whose goals were not. Within the Social Democratic debate, nationalism arguably took the form of what Elvander calls "welfare nationalism". This was a multi-faceted phenomenon. It was observable, for example, in the way some anti-EU campaigners criticised the conservatism, particularly regarding women's rights, that was said to dominate the Union's big Catholic member states, and which contrasted with the egalitarian political culture in Scandinavia. Lena Klevenås, a fierce opponent of accession from the beginning of the debate in SAP, was the loudest voice for these fears. She claimed that the pope had opposed the welfare state, and she raised fears of child-care responsibility being devolved from the state back to the family—and thus, most likely, the mother—under the Catholic principle of subsidiarity. She speculated that Sweden, as an EU member, would have to change its liberal abortion laws, either because of instruction from the Union or through necessity, as women from less enlightened European countries flocked north to obtain

---

[1] Evidence from all these sources is essentially qualitative and inductive. For this reason, I have not referenced every quotation made by an interviewee. This section claims only to be indicative; the identity of the interviewee was not the important factor, merely that his or her sentiment was expressed and was (arbitrarily) considered significant at this initial stage of analysis. Thus, it would hardly be worth interrupting the flow of the text with superfluous footnotes. Where a comment or a quotation has no specific reference attached, it may be assumed to have been recorded during interview.

terminations. Nor was she above making sweeping generalisations about religion and political culture:

> In the Protestant part of Europe, we formulate goals that are possible to attain and that we think are practically attainable. For a Catholic, it is the nature of things that one cannot reach goals. They are more 'aims', something to strive for. In the Catholic countries, people are expected to sin. And to receive forgiveness.[2]

Klevenås also stressed the point that the nation state, which had served Sweden so well, has a much lower priority in the Catholic countries.

Doubtless with such arguments in mind, Elvander saw Swedish welfare nationalism as an implicit expression of cultural superiority. He argued that whereas Swedish nationalism before the first world war had been of the right-wing authoritarian kind, afterwards the concept had been appropriated—as we saw in the previous chapter—by the politically domi-nant left, with the welfare state cited as proof that "Sweden is best; our country can be a pattern for other, less developed countries; Swedishness must protected from contamination by alien cultures."[3] Sweden, some in SAP argued, was a stable, developed, long-established nation state, with a successful and vibrant democratic culture. Countries without these attrib-utes might wish to pass more government functions to non-national levels, but that was not necessary for Swedes. Freedom of action outside the Un-ion would be a better means of advancing Social Democratic goals than working for them inside it.

"Isolationist" welfare nationalists argued that, from a national point of view, Denmark's experience with the Maastricht treaty illustrated how limited small countries' power was in the EU: legally, the treaty should have fallen when the Danes had said No in their first referendum, but their gov-ernment had been forced by the other member states to put the question to the electorate a second time. From a political perspective, it was claimed that while the conditions for the left's political dominance existed in Sweden, they were not present at the European level. Gröning, who edited an anthol-ogy of Eurosceptical Social Democrats' arguments, insisted that the powerful institutions in the Union—the Commission, the Council, the Court, the pro-posed European Central Bank (ECB)—were largely free from democratic

---

[2] Lena Klevenås, "Den katolska kyrkan och EG", in Lotta Gröning (ed.), *Det nya riket? 24 kritiska röster om Europa-Unionen* (Stockholm, Tidens förlag, 1993), p.88.
[3] Elvander (1994).

control. More importantly, the European labour movement, unlike the Swedish movement, was much too weak to sustain the left in holding government office.[4] A special motion to the 1993 party congress summarised the argument. "We have difficulty in believing in the idea of social democratic hegemony [in the EU]", its signatories declared.

> Social Democrats in Sweden and the Nordic region build chiefly on a people's movement...The several bourgeois parties in Sweden see themselves as idea parties. They do not seek sympathisers; rather, sympathisers seek the party. This is true of similar parties in Europe and, unfortunately, also the numerous left-wing parties. Many of the socialist parties are so-called 'lawyer parties'. They are, as a rule, elite parties, their efforts aimed at the highly educated. The poorly educated and the workers stand almost completely outside these parties...
>
> [T]here is a big difference in everyday policy between SAP in Sweden and, for example, the PSDI or PSI in Italy, Pasok in Greece, the PS in France, etc. The view of politics, democracy, employment, environmental protection, etc varies between Europe's social democrats...To believe that we Swedish Social Democrats can, like iron men, enter the European Union and bring about the political changes we desire, is simply pathetic.[5]

The chair of Social Democrats Against the EU, Sten Johansson, declared that the EU's social dimension, "this little opening", was scarcely enough to offset the power of the free market. A TCO economist called the social charter "a gesture to the gallery".[6] "For every Swedish Social Democrat", observed Walter Korpi, an eminent sociologist at Stockholm University, "there are more than 50 bourgeois votes in the EC."[7] A motion to the 1993 party congress argued that in pursuing Social Democracy's mission to build the good society, "It is not necessary to convince 12 other [EC] countries first—we can start sooner than that."[8]

In fact, as we saw, Elvander suggested that within SAP the pro-membership side was just as welfare nationalist as the anti-membership

---

[4] Lotta Gröning, "En nationalism som är internationell!", in Lotta Gröning (ed.), *Det nya riket? 24 kritiska röster om Europa-Unionen* (Stockholm, Tidens förlag, 1993).

[5] SAP, Uppsala branch, special motion 481 to congress, 1993.

[6] Roland Spånt, "SAP—ett slag i luften!", in Lotta Gröning (ed.), *Det nya riket? 24 kritiska röster om Europa-Unionen* (Stockholm, Tidens förlag, 1993), p.137.

[7] Walter Korpi, "Medlemskap = Massarbetslöshet", in Lotta Gröning (ed.), *Det nya riket? 24 kritiska röster om Europa-Unionen* (Stockholm, Tidens förlag, 1993), p.104.

[8] Stockholm branch, special motion 480 to congress,1993.

camp. Whereas the No-sayers were imbued with an "isolationist" variant of welfare nationalism, the Yes-sayers displayed a "missionary" type, in which "Sweden is a model that can be offered to EU countries, instead of being protected through isolation."[9] Although perhaps Elvander overstated his case, there was more than a grain of truth in it. "The Nordic model of society is perhaps the most humane the world has ever seen," Ingvar Carlsson, the party leader from 1986-96, insisted.[10] Ines Uusmann, chair of Social Democrats For the EU, averred that "In Sweden women have more power than probably anywhere else in the world, with the possible exception of Norway."[11]

Elvander's two types of welfare nationalism share pride and faith in Swedish Social Democratic achievements. But one type argues that it is sufficiently successful to be ripe for export; the other fears contamination and failure if those achievements are exposed to the EU. Thus we can say that there is enough evidence of a defensive, isolationist welfare nationalism being a significant cause of opposition to EU membership in SAP to propose, as our first hypothesised opinion cleavage, that a split between nationalists and internationalists lay behind the party's internal division.

Hypothesis 2: left v right

To the casual observer of the European issue in Nordic politics, there would seem to be a reasonably clear division in the respective party systems between left and right. The main parties of the right (except the Icelandic Independence Party) and the liberal parties (except Norway's) favour integration; social democrats are divided; and radical left parties are (with the partial exception of the Finnish Left Alliance) sceptical.

Yet it is important to clarify precisely what we mean by left and right. Swedish politics in general took a marked step to the left in the 1960s and 1970s, when LO's plans for the wage-earner funds were adopted by SAP. Perhaps just as significantly, the Centre Party in particular exploited a popular sense of alienation from political decision-making, which was perceived as being the preserve of the big corporatist organisations.

---

9 Elvander (1994).
10 Carlsson (1994), p.21.
11 Ines Uusmann, "Kvinnorna är halva EU—svenska kvinnors liv i morgondagens Europa", in Rolf Edberg and Ranveig Jacobsson (eds), *På tröskeln till EU* (Stockholm, Tidens förlag, 1992), p. 45.

The radical left sought to exploit this trend by applying a Marxian critique to the bastions of the traditional Social Democratic system—collective wage bargaining and the universal state welfare—on the grounds of their being, respectively, class collaboration and a bourgeois sop to the working class. If there was a modern trend in Swedish Social Democracy, the radical left appeared to be it.

Since then, however, the left in the Nordic region has changed, as it has elsewhere in Europe. The radical left parties no longer call themselves communist, though they remain generally Eurosceptical; and if there are radical, reformist elements within the social democratic parties, they are now inclined much more towards liberalism than towards Marxism. Undoubtedly, this development has many causes, including the collapse of East European communism. The changing social composition of social democracy's natural political constituency may also have played a part in shifting the party's centre of gravity towards the centre of the political spectrum.[12] But whatever the causes, it means that if we speak of "traditionalists" in SAP, we now refer to the left of the party, and "modernisers" denote those on the right who are prepared to contemplate policies such as lower taxes and retrenchment of welfare entitlements.

In the rest of Europe, left-wing critics of the EU usually focus on its removing state barriers to trade between the member states, and thus reducing government influence over the economy. More recently, however, and particularly in the Swedish context, in which the left has mostly had a much more benign view of free trade than it has elsewhere, the plans for EMU have become the foremost target of Eurosceptics' criticism. This was for three main reasons. First, there was the proposed constitution ECB, which prescribed its responsibility as fighting inflation above all else. Second, the bank was to be independent of political control. Third, the "convergence criteria" stipulated in the treaty would greatly limit scope for an expansionist fiscal policy. All these conditions—above all, the "Friedmanite" character of the ECB, as one anti-membership trade-unionist put it—were in conflict with Swedish tradition, in which fiscal and monetary policy had frequently been used to manage demand, with the priority of keeping down unemployment

---

[12] Odd Engström, an influential figure in SAP's upper echelons, acknowledged in an interview during the Social Democratic government's most difficult times in mid-1990 that he could see SAP becoming a middle-class party, because most of its membership was now middle-class—even if the party's policies would remain "classically Social Democratic". *LO-tidningen* May 19th 1990.

rather than inflation.[13] Some argued that, with Swedish governments already conforming to the EU's anti-inflationary monetary orthodoxy, the threat to the Swedish model from joining the Union was obvious. Meidner argued: "It is scarcely a coincidence that unemployment in Sweden has reached the EC's high levels at the same time as Sweden's powerful elites are preparing for EC membership."[14]

Even supporters of membership in SAP were generally less than enthusiastic about EMU, though some could see the advantages of being able to combat currency speculators more effectively, and others could accept a monetary policy that had price stability as its main goal. Allan Larsson, a former finance minister and a strong supporter of EU membership, claimed that European social democracy "aims at a simultaneous struggle against unemployment and inflation. We refuse to set one against the other. We have higher ambitions for the economy."[15] The general line of many pro-membership Social Democrats was that, in principle, EMU might have some advantages (reduced transaction costs, less risk for investors, a greater bulwark against currency speculators), but that there would also be costs, both economic and political. Yet many felt that, in the light of developments since Maastricht, EMU was unlikely to happen anyway. In any case, once inside the Union, Sweden might take steps to amend the convergence criteria and correct the lack of democratic control over the ECB. Social Democratic opponents of Swedish membership, on the other hand, rejected such hopes as naive, given the reality of power structures within the Union. Nor was an EMU opt-out clause, as obtained by Britain and Denmark, available to Sweden.

When asked directly whether the division over Europe in SAP reflected a simple left-right cleavage, interviewees' responses were often that No-sayers *did* tend to be more left-wing. Sten Johansson, the chair of Social Democrats Against the EU, agreed that it was "probably true". Certainly,

---

13 Economic arguments against EMU, based chiefly on the premise that a single monetary policy for economically different areas would result in economic stagnation and mass unemployment in less efficient regions, are expounded by, for example, the chief economist of the white-collar trade union organisation, TCO. See Roland Spånt, "Jobban offras för stabila priser", *LO-tidningen* January 29th 1993.

14 Rudolf Meidner, "Neutralitet och fullsysselsättning omodernt i ett EG-anslutet Sverige", in Lotta Gröning (ed.), *Det nya riket? 24 kritiska röster om Europa-Unionen* (Stockholm, Tidens förlag, 1993), p.104.

15 Allan Larsson, "Sätt Europa i arbete!", in Rolf Edberg and Ranveig Jacobsen (eds), *På tröskeln till EU* (Stockholm, Tidens förlag, 1994), p.37.

most were prepared to argue for a more reflationary, Keynesian economic policy than the current Swedish government (let alone the ECB) would consider, and were probably less prepared to countenance radical reappraisal of the way the Swedish welfare system operated, particularly regarding the size of the public sector and the weight of the tax burden.[16] But the picture was more complicated than that. Several of the most enthusiastic supporters of Swedish membership among interviewees said they considered themselves to be on the left, or even the radical left, of the party. Conversely, Johansson said he had always considered himself a "grey, in the centre": for nuclear power, against nuclear weapons; for collective capital-formation, against wage-earner funds.

Nonetheless, a second hypothesis is that the division in SAP over Europe reflected an ideological difference between the traditionalist left of the party and its modernising right.

Hypothesis 3: neutralists v supporters of collective security

The pro-membership campaigners in SAP generally had a pragmatic view of Swedish security policy. To them, neutrality had always been contingent upon geopolitical circumstances in Europe, and the cold war had created conditions in which a relatively strict interpretation of the policy had been required. It had been (in one interviewee's words) "a means, not an end", and could now be modified. Thus, the strict consistency and credibility of policy that had previously been essential—a key component of which was an independent trade-policy—could be relaxed in the 1990s. Moreover, the EU could be seen as a "peace project". As we saw in chapter 3, collective-security schemes, whether regional (based on the Nordic region) or, especially, global (based on the League of Nations and the United Nations), have long held an attraction for some Social Democrats.

SAP's No-sayers, in contrast, had a very different view of Swedish neutrality. Far from being a product of the cold war, it was seen as having a much longer tradition. Johansson considered it bound up with Swedish independence and democracy; he even talked, somewhat dramatically, in terms of the survival of a small people. That was an example of a Eurosceptical defence of neutrality on principle. A more pragmatic defence, also expounded by Johansson, was that relying on the West for support against Russia was a dangerous strategy for Sweden (and Finland) to pursue, in the light

---

[16] See Wibe (1993).

of historical experience. In such circumstances, neutrality gave Sweden a weightier voice in the world than it could ever have in collective European security arrangements. For SAP's Eurosceptics, then, neutrality's salience survived the cold war. Johansson acknowledged that the Maastricht treaty had established foreign policy co-ordination on the basis of unanimity, but pointed both to treaty protocols requiring member states to desist from blocking common foreign-policy positions, and to Sweden's promise at the outset of the entry negotiations not to obstruct moves towards common defence; both were threats to Swedish neutrality, he argued. He even claimed that British and French nuclear weapons would be an integral part of a common EU foreign policy.[17]

Our third hypothesis, then, is that a long-standing division over Sweden's stance on the international stage, between neutralists and those preferring collective security arrangements in co-operation with other states, was reflected in the division over the European issue.

Hypothesis 4: free-traders v pragmatists

In other European countries, it has usually been the supporters of free trade who have been in favour of European integration and the opponents who have been more sceptical. This has been clearly demonstrated during the Eurosceptical stages that the French Socialists and Communists passed through (and, in the latter's case, may still be passing through), in the British Labour Party's radical-left phase during the 1980s and, perhaps, in the Greek Socialists' more nationalist stages. In Sweden, however, there is little evidence of a correlation between left-wing radicalism and protectionism of the type that would be likely to result in a Eurosceptical policy position. Indeed, "the principle of free trade is virtually unquestioned in Swedish public debate."[18]

In fact, there is arguably more evidence that some Swedish Social Democrats opposed EU membership on the grounds that the Union was *not sufficiently* committed to free trade. We saw how and why Sweden's labour movement became reconciled to free trade at an early stage of its history. In part, this was unavoidable, given the country's size and necessary dependence on exports for its prosperity. But, as we saw in the previ-

---

17 Sten Johansson and Maj Britt Theorin, "Strategigrupp utreder kärnvapens roll i EU", *Svenska Dagbladet* November 7th 1994.
18 Sjöstedt (1987), p.5.

ous chapter, the promotion of free trade also found favour with Social Democrats who wanted to show solidarity with the poor countries of the world—a significant tradition in the party.[19] Sweden could justly claim to have opened its markets to cheap produce from the third world. The EU, in contrast, was far less accommodating to poorer countries. The common agricultural policy (CAP)—with its barriers to imported agricultural produce and, worse, its export subsidies that served to dump surpluses on the world market—was only the worst example of its attitude.

One Social Democratic activist argued that "This [the EU] is not a socialist community." In the next breath he defended free trade; argued that membership would force Sweden to raise tariffs against the rest of the world; criticised the EU for its predilection for economic planning, as epitomised by the CAP; and insisted that social democracy should not try to order the market, but rather to redistribute its fruits through high taxes and welfare benefits. Johansson expressed similar sentiments. "Free trade with other countries can…be characterised as a necessary prerequisite for high living-standards in a country," he argued. "If we…are to have the chance of saving the welfare state, it means keeping as much trade freedom as possible."[20]

Although the EEA was considered sufficient for Sweden's economic interests by some pro-membership Social Democrats, others felt that it was not enough. The chair of the Paper-Workers' Union, for example, feared that non-tariff barriers, such as custom controls at EU borders, could still disadvantage Sweden if it stayed outside the Union, and that higher interest rates might jeopardise much-needed investment.[21] By contrast, the EEA was seen by most leading Eurosceptical Social Democrats as, at least, a reasonable deal for Sweden, giving it the market access it required but avoiding involvement in the common trade policy, the CAP and EMU. In interview, Johansson praised the terms of the EEA as providing for free trade between Sweden and the EU while also having potential for developing certain "flanking policies", including a social dimen-

[19] Cf. Ole Elgström, "Socialdemokratin och det internationella solidaritet", in Bo Huldt and Klaus Misgeld (eds), *Socialdemokratin och svensk utrikespolitik. Från Branting till Palme* (Stockholm, Utrikespolitiska institutet, 1990).

[20] Sten Johansson, "Löntagarstrategier inför kapitalets internationalisering", in Lotta Gröning (ed.), *Det nya riket? 24 kritiska röster om Europa-Unionen* (Stockholm, Tidens förlag, 1993), p.72, 83.

[21] Sune Ekbåge, "Jobben, lönerna och exportberoendet", in Rolf Edberg and Ranveig Jacobsson (eds), *På tröskeln till EU* (Stockholm, Tidens förlag, 1992), pp.60-63.

sion. For him, an EEA that added controls over currency movements would be his "Utopia". In addition, an anti-EU trade-union activist described himself in interview as on the radical left, but also radically in favour of free trade, the market-led shift to a service-based economy, and even the need to attract international capital. The EU's regulation and bureaucracy, he argued, would only put investors off Sweden.

Therefore, our fourth hypothesis is that, to many Social Democrats, the EU represented a violation of the party's traditional acceptance of—indeed, enthusiasm for—free trade, being seen as a protectionist, mercantilist organisation. This was reflected in opposition to EU membership. Supporters of accession, meanwhile, while certainly not anti-free-trade, took a rather more pragmatic approach to the issue. Support for free trade could be tempered by other advantages, political and economic, that joining the Union would confer.

Hypothesis 5: greens v greys

There was surprisingly little observable evidence in the referendum campaign of a division between Social Democrats who prioritised economic growth, and those who prioritised non-materialist objectives, such as a clean environment. No interviewee gave the environment as his or her main reason for voting for or against accession. Yet there is an a priori case for environmentalism being an important element in the modern Social Democratic Party. In *The Transformation of European Social Democracy*, Kitschelt cites SAP as an example of a party that finds itself courting an electorate with a high degree of post-materialist preferences, and which has adapted to these circumstances relatively well. It may also be that it played some role in the party's European debate.

The environment is clearly an issue in Swedish politics, but it is not always expressed in inter-party cleavages. The Greens were formed as a party in 1981; they first entered the Riksdag in 1988 and after 1994 appeared to establish themselves as a parliamentary force. Previously, however, if a cleavage between materialists and post-materialists has existed, it has been overlaid by other political cleavages, and utilised—with, it would seem, divergent fortunes—by the Left and Centre parties. Some observers suggested that, in fact, the "new" cleavage actually represented a reactivation of one of those "old" cleavages identified by Lipset and Rokkan, that between rural and urban economic interests. However, Bennulf argues that there is little in existing Swedish survey data to suggest the existence of a cleavage that is distinct from left–right, the only exception being over nuclear power.[22] In sum, there is little evidence that a significant materialism–post-materialism dimension exists in Swedish party politics, or even in the Swedish electorate.

Yet within SAP there is, nevertheless, evidence of a significant post-materialist lobby, which prioritises environmentalism over economic growth. Differences between the "green" and "grey" wings of the party found expression in argument about plans for a bridge over the Öresund between southern Sweden and Denmark. It was the subject of particularly bitter argument at the party congress in September 1990, where the communications minister succeeded in attaining only conditional approval for the project, and even that was in the face of strong opposition from SSU, the Federation of Social Democratic Women and Christian Social Democrats. They argued that the environmental impact on the area—especially if, as the Danes were insisting, the bridge would carry only road traffic rather than trains—would be too great. (Eventually, the Social Democratic government agreed in spring 1991 to support the bridge.)

A persistent and more damaging row has occurred over nuclear power. A referendum in 1980 on Sweden's nuclear programme, held in the aftermath of the accident at Three Mile Island, found a plurality in favour of phasing it out by 2010. SAP and LO, many of whose members were concerned at the consequences for employment of abandoning nuclear energy, backed this line, but it only just defeated another of the three options on offer to voters, that of closing the reactors forthwith. A good number of Social Democrats preferred this policy. Polls suggested that only about two-thirds of the party's supporters backed its leadership's line, despite its mounting a strong campaign for a "cautious decommissioning". The issue was revived by the Chernobyl nuclear disaster in 1986, and once more it caused serious friction within SAP. In 1990, authority was given to negotiate with other parties on the manner of the decommissioning, although this was still to be achieved by 2010.

A vague agreement for a five-year energy programme was agreed with the Centre and the Liberals in January 1991. As Sweden's economic situation deteriorated, it seemed increasingly likely that the party might take a more pragmatic stance, and lay more emphasis on the economic

---

22 Martin Bennulf, "En grön dimension bland svenska väljare?", *Statsvetenskaplig Tidskrift* vol. 95, 1992, pp.329-58.

costs of decommissioning. Yet Carlsson, for one, insisted that the party remained bound by the referendum's verdict of 1980; and after formal parliamentary co-operation with the Centre was established in April 1995, the pressure on the party leadership to continue the process increased. In February 1997 the Social Democratic government incurred strong criticism from both business and trade unions for announcing the closure of two reactors, one by mid-1998 and the other by 2001, before their useful life had been completed—the first such decision in the world. Even then, however, the future of Sweden's other nuclear power plants, which then provided more than half the country's electricity, remained uncertain.

Given SAP's recent history, it would be a surprise if there were no difference of opinion between "grey" and "green" Social Democrats on Europe. The Stockholm branch of SSU, in a motion to the 1993 party congress, argued:

> It is impossible to achieve equality by lifting the whole world up to our standard of living. If we are to achieve equality, the rich world must reduce its consumption. Yet despite this, growth remains the goal, growth that in today's form increases use of resources and creates still more pollution. Demand for growth in the rich world is a motor of environmental damage and makes equality between north and south impossible.
>
> The course the EC has set out has one overriding goal: to increase economic growth, and to make the EC even richer...It is impossible to implement an effective defence of the environment when free competition and economic growth are made the overriding objectives...
>
> At the same time the EC prevents countries from taking the necessary steps away from this negative development.[23]

In contrast, some leading supporters of membership, particularly in the trade unions, made it clear that their objective was to achieve precisely this sort of economic growth for Sweden.

In any case, the fact that some categories of people—younger voters, for example—were likelier to vote for the environmentalist Left and Green parties, and were also more likely to have voted No in the EU referendum, gives us some basis for making our fifth hypothesis. It is that there was a fault line in the party that divided those Social Democrats with materialist objectives, such as economic growth, and those with non-materialist objectives, such as protection of the environment. This cleavage was re-

---

[23] SAP, Stockholm branch, special motion 480 to congress, 1993.

flected in the division over the EU: materialists voted Yes, because of the promise of higher growth for Sweden in the Union; and non-materialists voted No, because of a wish to preserve Swedish environmental standards and because the EU was seen as being inherently favourable to growth at the expense of environmentalism.

Hypothesis 6: influencers v democrats

Some Social Democratic Yes-sayers insisted that EC membership was a question of democracy. Carlsson, for example, claimed: "Democracy is more than discussing or negotiating. It is also being able to make decisions that matter."[24] A contributor to the debate about the party's ideological direction put the argument succinctly.

> A state like Sweden is formally sovereign, but today this sovereignty is in practice redundant...[M]odern society's development has upset the symmetry between influencers and the influenced, between those who take decisions and those affected. This relationship no longer applies within the border of the nation state. The borders between domestic and foreign policy have, quite simply, been rubbed out.
>
> Thus, I see the EU as an attempt to create a new symmetry—a new order for decision-making in cross-border social questions.[25]

Such views are surely to confuse democracy with power. Democracy is about the way in which decisions are made; power concerns what those decisions are about. The more thoughtful supporters of membership acknowledged that there might be a price to be paid in democracy for Swedish membership. But they stressed that for any country there has to be balance between democracy and efficiency, economic and political, and that joining the EU would strike the right balance for Sweden. In the words of one member of SAP's National Executive, democracy is not an end in itself, but has to be used to achieve something else. One senior figure in Social Democrats For the EU admitted that the EU was *not* a democratic polity, but that was not the point: it was an international organisation, not a state. Although the "democratic deficit" concerned him, the single market needed supranational rule-making, and the EC provided both it and the means to regulate transna-

---

24 Carlsson (1994), pp.20-21.
25 Helle Klein, "Demokrati—en fråga om inflytande och tillgänglighet", in Rolf Edberg and Ranveig Jacobsson (eds), *På tröskeln till EU* (Stockholm, Tidens förlag, 1992), p.61.

tional markets—a fundamental social democratic objective. Supranational politics, Carlsson claimed, was needed to complement an internationalised economy, to "reassert the balance between the social economy and the business economy...[T]he European left needs common democratic institutions so that our political ambitions can be realised." Comparing the EU with the EEA, he declared: "It is sometimes suggested that the EEA...is enough for Sweden. If I was a director of Stockholm's chamber of commerce I would agree. But for me as a Social Democrat it is completely inadequate."[26]

For many No-sayers, however, the terms of this trade-off between democracy and influence were unacceptable. Theirs was an opposition on principle to joining to a supranational polity. To Johansson, for example, democracy—or perhaps, more precisely, the Swedish notion of *folkstyre* (literally, people's steering)—was an end in itself. A functioning democratic society, he argued, required the mechanisms of both "vertical and horizontal communication", by which he meant, *inter alia*, a common language, culture and media.[27] The EU could never hope to attain this; its undemocratic institutions were all at odds with Swedish democratic tradition. This was his fundamental objection to EU membership. It can be characterised as a purposive nationalism, with the nation state seen as a means for achieving another goal—namely, democracy. A senior member of the Social Democratic pro-membership side agreed that the most perceptive and persuasive members of the No lobby—people like Johansson and Marita Ulvskog, Carlsson's former press secretary and by then editor of a newspaper in northern Sweden, *Dala-demokraten*—did not stress the economics of the issue, since Sweden's integration with the European economy, consummated in the EEA, was already a fact. Instead, they emphasised non-economic objections to Swedish membership: democracy, sovereignty, internationalism.

Our sixth and final hypothesis, therefore, is that the division between opponents and supporters of EU membership in SAP reflected a disagreement over the relative merits of Swedish democracy and Swedish influence. For the pro-EU side, expanding national influence was to be prioritised in this particular situation. For the anti-EU side, either the gain in influence did not compensate for the loss of democracy, or it *could* not compensate, so important was the principle of *folkstyre*; either way, in this particular situation democracy was to be prioritised over influence.

---

[26] Carlsson (1994), pp.19-20.
[27] Such arguments are also prominent in Danish Eurosceptical circles. See, for example, Junibevægelsen, *Programme of the JuneMovement* (Junibevægelsen, Copenhagen, 1994).

Table 4.1 Motions on EU membership submitted to SAP's 1993 congress

Motion	Local party	Urging Yes	Uncommitted	Urging No
465	Uppsala	s		
466	Stockholm			s
467	Gotland			s
468	Stockholm			s
469	Nacka			s
470	Lund			s
471	Örebro			s
472	Solna	s		
473	Norrköping	s		
474	Vindeln			n
475	Gagnef			s
476	Linköping	n		
477	Lund	s		
478	Älmhult		n	
479	Sollentuna	s		
480	Stockholm			s
481	Uppsala			s
482	Örnsköldsvik			n
483	Umeå			s
484	Ljusdal		n	
485	Lund	s		
486	Stockholm		s	
487	Luleå			n
488	Alingsås		n	
489	Lund		n	
490	Luleå			n
491	Falköping		n	
492	Karlstad		n	
493	Värmdo		n	
494	Trollhättan		n	
495	Stockholm		s	
496	Lindesberg		s	
497	Jönköping		n	
498	Malmö	s		

Note: n = normal motion, s = special motion (ie, not adopted by local party).
Source: Congress protocols.

Figure 4.1 Classification of pro- and anti-membership arguments in motions put to SAP congress, September 1993

Fault line	Anti-EC arguments	Pro-EC arguments
"Isolationists" v "missionaries"	Little in common with continental left (480-83)	Sweden's European cultural identity (465, 477)
	Alternative of closer Nordic co-operation (481-83) Threat to Swedish drugs, alcohol, consumer-protection policies (480-83)	No Nordic alternative (479)
	Capitalism world-wide, not just EU-wide (480)	EC as means to control international capitalism (465, 473, 477, 479)
	Swedish economy can be influenced (480) More influence outside, demonstrating alternatives to EC (480-83) Sweden spared EU contributions (468, 469, 470, 471,)	EC provides means for Keynesian demand-management (472)
Left v right	Threat to "Swedish model", "good society" (tax, thus/and welfare, policies), danger of "two-thirds society" (467-68, 474, 480-83)	Economic necessity of maintaining welfare resources (472, 476)
	EC as neo-liberal, with inadequate social dimension: threat to Swedish worker-protection, trade unions, social dumping (480) EMU would mean permanently high unemployment (468, 480-83)	EC has its social dimension (485)

Neutralists v collective-security	Threat to nonalignment in peace, aiming at neutrality in war; non-EC common-security alternatives (467-71, 480-83, 490) No means of leaving EC once in (481-83)	Nonalignment possible within EU (485) Security inside EC; neutrality no longer appropriate (472, 479)
Free-traders v pragmatists	Greater scope for solidarity with poor countries, including free trade (467-71, 475, 480-83) EEA provides economic benefits (468-71, 480-83)	EC can be a platform to promote East European interests (476)
Greens v greys	EU's inferior environmentalism, growth/liberal-orientation (480-83)	EC has its green potential, especially post-Maastricht (485) EC sometimes has better consumer-protection (485)
Democracy v influence	EC decision-making incompatible with democracy (468-71, 480-83)	EC can be democratic (465, 485) Democracy balanced by efficiency (485) EMU needed to combat speculators (479, 485) EMU can balance Bundesbank influence (472) EEA provides no political input; full membership does (485) International co-operation against crime needed (485) International solutions to international problems (485)

Source: Congress protocols.

Motions submitted to the 1993 party congress

Can our hypothesised ideological cleavages within SAP find any initial evidence to support their existence, before we test them against our questionnaire survey? One option is to analyse the content of motions submitted to the party congress held in Gothenburg in September 1993. Motions are submitted by individual members first to a SAP local branch, which either adopts it, rejects it or agrees to submit it to congress as a "special" motion, without adopting it. A total of 34 motions that dealt with the EU were submitted to congress in 1993 (see table 4.1). Of these, eight urged it to support Sweden's accession, 14 urged it to oppose it and 16 were uncommitted (most arguing that it was too soon to take a decision). The division in the party can thus be gauged even at this stage. The question most relevant in this context, however, is what arguments each side was using.

Figure 4.1 depicts the arguments employed about whether Sweden should or should not join the EC. The classification of the arguments using the six hypotheses proposed in the previous section can be seen to capture them quite impressively. Of course, as with the inductive method used originally to generate the hypotheses, there must be an element of arbitrariness and personal judgment involved in the classification. For instance, it could be argued that the distinction between, on one hand, the fault line between welfare nationalists, and, on the other, that between left and right, is a debatable one. It is certainly the case that the arguments in each of these categories are very similar. Yet there was a difference. In the latter, there was a disagreement about goals, about Sweden's becoming more like the countries of the EC, and less like the traditional Swedish model of Social Democracy. In the former category, there was essentially a disagreement about means, about whether shared objectives were better pursued inside or outside the Community.

It is now appropriate to test these hypotheses more systematically, using the questionnaire data.

Testing the hypotheses

The first part of our data analysis concerns how Social Democratic activists in Sweden intended to vote in the referendum when the questionnaire was sent to them, four months before it took place.

Figure 4.2   Social Democratic activists' voting intention (June 1994) in the EU referendum

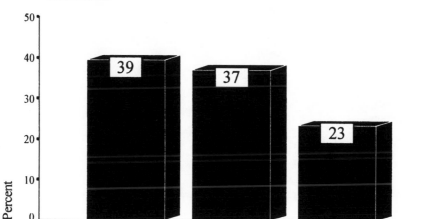

How will you vote in the EU referendum?

This is a fair reflection of the way opinion polls in summer 1994 were suggesting the Swedish electorate as a whole was inclined to vote. The large number of undecided SAP activists indicated in figure 4.2 certainly accords with what the polls were indicating about voters' general uncertainty, although the proportion of Social Democratic voters that were thought likely to vote Yes was rather smaller than that found here for the party's activists, perhaps indicating the greater sense of loyalty party activists had towards their leadership (and its pro-membership position).

Hypothesis 1: "isolationists" v "missionaries"

When asked what "Social Democratic opponents of EU membership are primarily motivated by", nearly a fifth (19.1 per cent) said it was nationalism. Table 4.2 shows the responses to questions designed to test both the level and—following Elvander's distinction between isolationist and missionary—the type of Swedish welfare nationalism among Social Democratic activists.

Table 4.2 Testing the welfare nationalist fault-line

Referendum voting intention

	Yes %	Uncertain %	No %	total %
How often do you feel "European"?				
Never	10.8	14.1	32.4	19.5 (N = 59)
Seldom	15.0	26.8	24.3	21.2 (N = 64)
Sometimes	48.3	39.4	27.0	38.4 (N = 116)
Often	25.8	19.7	16.2	20.9 (N = 63)
total	100 (N = 120)	100 (N = 71)	100 (N = 111)	100 (N = 302)

"Continental Europe's Catholic tradition would pose problems for Sweden as an EU member"

	Yes %	Uncertain %	No %	total %
Fully agree	4.1	7.0	16.1	9.2 (N = 28)
Partly agree	19.0	19.7	32.1	24.0 (N = 73)
Uncertain	16.5	33.8	27.7	24.7 (N = 75)
Disagree	60.3	39.4	24.1	42.1 (N = 128)
total	100 (N = 121)	100 (N = 71)	100 (N = 112)	100 (N = 304)

"National sovereignty is essential to preserve Sweden's national identity"

	Yes %	Uncertain %	No %	total %
Fully agree	31.1	30.0	49.5	37.7 (N = 113)
Partly agree	38.7	52.9	32.4	39.7 (N = 119)
Uncertain	6.7	8.6	5.4	6.7 (N = 20)
Disagree	23.5	8.6	12.6	16.0 (N = 48)
total	100 (N = 119)	100 (N = 70)	100 (N = 111)	100 (N = 300)

"Sweden's economy can still be effectively influenced by Swedish politicians"

	Yes %	Uncertain %	No %	total %
Fully agree	37.0	22.5	37.7	33.9 (N = 103)
Partly agree	53.8	59.2	42.1	50.7 (N = 154)
Uncertain	2.5	5.6	4.4	3.9 (N = 12)
Disagree	6.7	12.7	15.8	11.5 (N = 35)
total	100 (N = 119)	100 (N = 71)	100 (N = 114)	100 (N = 304)

The figures illustrate a division within the party's activists over Sweden's relationship with the rest of Europe. Four in ten respondents (40.7 per cent) said they never or only seldom felt "European", but nearly six in ten (59.3 per cent) said that sometimes or often they did. Yet when these figures are controlled for voting intention in the referendum, the results are only moderately indicative of a correlation between this and "nationalist" or "Europeanist" sentiments. A third (32.4 per cent) of those intending to vote No said they never felt European; but so did over a tenth (10.8 per cent) of those intending to vote Yes. Nearly a quarter of those who intended to vote No (24.3 per cent) seldom felt European; but so did 15 per cent of those intending to vote Yes. Conversely, 43.2 per cent of those who intended to vote No sometimes or often felt European. In other words, there is a correlation between feeling a European identity and voting Yes to EU membership, but it is a relatively weak one, with plenty of respondents bucking the trend.

Two questions in this category did not hint at division between SAP's two types of welfare nationalists. Responses to the statement, "National sovereignty is essential to preserve Sweden's national identity," was designed as a straightforward attempt to test nationalist sentiment, and Eurosceptics were expected to be more in agreement with the statement than those intending to vote for EU membership. Yet there was no very strong correlation. Over three-quarters of all respondents (77.4 per cent) either fully or partly agreed with the statement. Clearly, the concepts of national sovereignty and Swedish identity were not something the large majority of Social Democratic activists were seeking to discard, whether they were for or against EU membership.

Finally, even more surprising responses can be seen to the statement, "Sweden's economy can still be effectively influenced by Swedish politicians." It was expected that isolationist welfare nationalists would be confident that the economy had not become so prey to international forces that national government had been made powerless to manage it, and that they would thus be less inclined to see a need for EU membership. In contrast, missionary welfare nationalists would perhaps see the EU as a platform from which the Swedish model could be taken into a new phase (or perhaps salvaged), or from where its characteristics could be exported to the wider continent. However, the proportions of respondents planning to vote Yes and No who fully agreed with the statement were nearly equal (37 per cent and 37.7 per cent respectively); and, of those who partly agreed, a bigger proportion planned to vote for the EU (53.8 per cent) than to vote against (42.1 per cent). This suggests that Social Democratic activists had not given up on their leaders' ability to steer the Swedish economy, and most did not see EU membership as a means to re-establish political control of the international market.

It seems, then, that probable No voters *were* more isolationist, in Elvander's terms, than probable Yes voters; but the difference was not very great.

Hypothesis 2: left v right

The questions in this section were intended to test if respondents who could be categorised as more left-wing took a more Eurosceptical approach than those identified as more right-wing. As we have seen in this chapter, the anecdotal evidence of this being an important reason for the division in the party over EU membership is quite strong. Moreover, nearly a third of respondents to the survey (29.1 per cent) suggested that a more left-wing ideology was the biggest reason why a Social Democrat should oppose Sweden's joining the Union. But could a more systematic analysis of opinion support this hypothesis?

In attempting to define left-wing and right-wing in the Swedish Social Democratic context, an obvious starting point is self-placement. Table 4.3 depicts answers to the simple question, "How would you categorise yourself politically in the party?" They show only the smallest correlation between responses and intention to vote in the EU referendum. True, of those likely to vote No, the proportion of those who placed themselves on the radical left was more than three times as large (15.7 per cent) as the proportion of probable Yes voters who categorised themselves similarly (5.1 per cent). But these were small numbers. The large majority of respondents, 71.3 per cent, placed themselves on the left of the party; and they made up a slightly bigger proportion of those who intended to vote No (72.2 per cent) than those intending to vote Yes (67.5 per cent). Self-placement on the left–right scale, then, offers little evidence of a correlation with being pro- or anti-EU.

Table 4.3 Testing the left v right fault-line

	Referendum voting intention			
	Yes %	Uncertain %	No %	total %

How would you categorise yourself politically within the party?

	Yes %	Uncertain %	No %	total %
Radical left	5.1	8.8	15.7	9.9 (N = 29)
Left	67.5	76.5	72.2	71.3 (N = 209)
Centrist	23.1	14.7	9.3	16.0 (N = 47)
Right	4.3	-	1.9	2.4 (N = 7)
total	100 (N = 117)	100 (N = 68)	100 (N = 108)	100 (N = 292)

"Sweden's high tax- and welfare-levels—the Swedish model—must be reduced to adapt to new circumstances."

	Yes %	Uncertain %	No %	total %
Fully agree	4.2	4.2	2.7	3.6 (N = 11)
Partly agree	23.3	22.5	24.3	23.5 (N = 71)
Uncertain	1.7	2.8	2.7	2.3 (N = 7)
Disagree	70.8	70.4	70.3	70.5 (N = 213)
total	100 (N = 120)	100 (N = 71)	100 (N = 111)	100 (N = 302)

How would you categorise the party leadership politically?

	Yes %	Uncertain %	No %	total %
Radical left	1.8	7.4	3.8	3.8 (N = 11)
Left	43.4	47.1	34.3	40.9 (N = 117)
Centrist	46.0	38.2	37.1	40.9 (N = 117)
Right	8.8	7.4	24.8	14.3 (N = 41)
total	100 (N = 113)	100 (N = 68)	100 (N = 105)	100 (N = 286)

In general terms, what is your perception of the EU politically?

	Yes %	Uncertain %	No %	total %
Left-wing/social democratic				
	21.3	4.8	1.9	10.1 (N = 28)
Centrist	1.9	3.2	3.8	2.9 (N = 8)
Open to change				
	50.0	71.0	86.8.0	27.2 (N = 75)
Right-wing/neo-liberal				
	26.9	21.0	7.5	59.8 (N = 165)
total	100 (N = 108)	100 (N = 62)	100 (N = 106)	100 (N = 276)

This impression is reinforced in one way by responses to the question, "In general terms, what is your perception of the EU politically?" Almost six out of ten respondents (59.8 per cent) thought that the Union, rather than being on the left, centre or right, was "open to change". But that figure contained exactly half of those respondents who were inclined to vote Yes, compared to barely a fifth of those inclined to vote No (21 per cent). Even more strikingly, whereas over a quarter of probable Yes voters considered the EU to be right-wing or neo-liberal (a notable enough figure in isolation), not much less than nine in ten of probable No voters (86.8 per cent) had this opinion of the EU.

On the other hand, table 4.3 also depicts the responses to the statement: "Sweden's high tax- and welfare-levels—the Swedish model—must be reduced to adapt to new circumstances." While many on the non-socialist side of Swedish politics might agree with such a statement, it could be expected that only a small number of ultra-reformers within SAP would accept its prescription. In fact, while a large majority of respondents (70.5 per cent) disagreed with the statement, an unexpectedly significant minority, 27.1 per cent, fully or partly agreed with it. This is, it would seem, evidence of the divide between modernisers and traditionalists within SAP. But, as far as attitudes to the EU are concerned, there is—arguably even more surprisingly—no link at all between being, on one hand, a Eurosceptic and a traditionalist, or between being, on the other, pro-EU and a moderniser. In fact, conversely, a bigger proportion of probable Yes voters (70.8 per cent) disagreed with the statement than probable No voters (70.3 per cent), while a bigger share of those intending to vote No (24.3 per cent) partly agreed with the statement than those planning to vote Yes (23.3 per cent).

In sum, evidence in support of the hypothesis that the divide within SAP over EU membership corresponds to that between left and right is present, but decidedly limited. There seems to be a perception among activists that a left–right fault line within the party was a factor in determining Social Democrats' stances towards the EU. But, when more objective measurements of this cleavage are attempted, the distinction between two ideological wings—or, more specifically, between traditionalists and modernisers—becomes hard to discern.

Hypothesis 3: neutralists v supporters of collective security

As we have seen, during the heyday of the Swedish model in the early 1970s neutrality came to be perceived by many Social Democratic activists as part of the country's national identity. Equally, however, the party also has a tradition of openness to the possibility of engagement in collective security arrangements. For those belonging to this latter tradition, neutrality was much more a pragmatic policy, which could be adapted according to circumstances. Thus, three possible responses to the question, "Which, if any, of the following statements is closest to your prescription for Swedish security policy?", were presented to the survey sample, each reflecting a different policy option.

The most immediately notable result, represented in table 4.4, is that just one in ten respondents (10 per cent) thought that neutrality should be abandoned, and that Sweden should embrace both the EU's common foreign and security policy and even common defence, an aspiration expressed in the Maastricht treaty. The rest of the respondents were divided almost equally between those wanting to maintain Swedish nonalignment, but within the EU (44.5 per cent), and those who wanted to retain the traditional Swedish security-policy formula of nonalignment in peace, aiming at neutrality in war, outside the EU (45.5 per cent). It is not surprising, though, that, of the 90 per cent of respondents who wanted to keep one version of nonalignment, the majority of those planning to vote Yes hoped to do so within the EU, and the majority of those planning to vote No hoped to do so outside. Indeed, rather more surprising is that 14.3 per cent of probable No voters wanted Sweden to remain nonaligned *within* the EU, and that 15.8 per cent of probable Yes voters wanted to stay nonaligned *outside* the EU. Interpretations of such seemingly illogical responses could range from simple confusion on the part of the respondents to a suggestion that, as in other policy areas, a good number planned to vote in the referendum despite, not because of, their sincere preferences in the field of Swedish security policy.

Table 4.4 Testing the neutrality v collective-security fault-line

Which, if any, of the following statements is closest to your prescription for Swedish security policy?

Referendum voting intention

	Yes %	Uncertain %	No %	total %
Sweden should be nonaligned in peace, aiming at neutrality in war, outside the EU				
	15.8	33.3	84.8	45.5 (N = 137)
Sweden should remain nonaligned within the EU				
	62.5	62.3	14.3	44.5 (N = 134)
Sweden should take full part in EU security policy, perhaps leading in time to common defence, as described in the Maastricht treaty				
	21.7	4.3	0.9	10.0 (N = 30)
total	100 (N = 120)	100 (N = 69)	100 (N = 112)	100 (N = 301)

This raises the basic problem with the data derived from this survey, which is particularly acute in this section. It is that responses can be tested for correlation, but not for causality. We can see that there is a connection between security-policy preferences among Social Democratic activists and their position on joining the EU. But we cannot tell if the former informed the latter, or vice versa.

Hypothesis 4: free-traders v pragmatists

Our hypothesis in this section was not an obvious one. It was that Swedish Social Democrats, far from being suspicious of free trade as many of their European counterparts historically had been, were in fact strongly committed to it. Moreover, rather than seeing European integration as a means of promoting free trade, people in SAP who opposed Sweden's joining the EU did so on the grounds that the Union inhibited free trade, especially with the poorer countries of the world. Four questions in the survey were designed to test whether this hypothesis had any explanatory worth—and, for that matter, whether the Swedish Social Democrats' reputation for adherence to free-trading principles was actually deserved.

The results from the first question in this section, "What is your general attitude to free trade?", suggest that, by and large, the reputation is deserved (see table 6.6). Over three-quarters of respondents (77.2 per cent) were in favour of free trade, which itself is a notable balance of opinion in a left-of-centre party. As for this issue's correlation with the European question, over nine out of ten probable Yes voters (91.7 per cent) supported free trade, compared to fewer than two-thirds (65.5 per cent) of probable No voters. The fact that so many Eurosceptics were in favour of free trade is, in a European context, striking. But it certainly does not bear out the suggestion that activists inclined to vote No to the EU were actually likely to be more supportive of free trade than those inclined to vote Yes. A similar inference can be drawn from the responses to questions pertaining to the EEA. Again, it is notable that of probable opponents of EU membership, 59.6 per cent were in favour of EEA membership. But it must also be acknowledged that, of probable Yes voters, the figure was fully 87.6 per cent.

Finally in this section, activists were asked if they thought there should be limits to free trade between Sweden and countries with lower social protection. This was designed to test commitment to the principles of free trade under less abstract circumstances, which might be expected to prick a social democratic conscience. Opposition to trading with countries and firms whose workers suffer, for instance, poor working conditions or very low rates of pay has been a feature of left-wing politics throughout the West for some time. On the other hand, because such conditions are largely inevitable in the poorest countries, precisely because they are poor, the counter-argument is that trade with firms and countries whose workers endure them is not only unavoidable but also desirable, as the wealth created will ultimately improve the lot of those workers. As one respondent wrote on the questionnaire, if Sweden were to impose restrictions on trade with countries that had lower social standards than its own, there would be none in the world with which it could actually trade freely.

Table 4.5 Testing the free-traders v pragmatists fault-line

Referendum voting intention

	Yes %	Uncertain %	No %	total %
What is your general attitude to free trade?				
For	91.7	71.4	65.5	77.2 (N = 234)
Uncertain	7.5	25.7	29.2	19.8 (N = 60)
Against	0.8	2.9	5.3	3.0 (N = 9)
total	100 (N = 120)	100 (N = 70)	100 (N = 113)	100 (N = 303)
What is your opinion of Sweden's membership of the EEA?				
For	87.6	67.6	59.6	72.5 (N = 222)
Uncertain	9.9	31.0	28.9	21.9 (N = 67)
Against	2.5	1.4	11.4	5.6 (N = 17)
total	100 (N = 121)	100 (N = 71)	100 (N = 114)	100 (N = 306)
Should there be limits to free trade with countries that have lower social protection than Sweden?				
Yes	42.9	56.1	57.7	51.4 (N = 152)
Uncertain	27.7	24.2	23.4	25.3 (N = 75)
No	29.4	19.7	18.9	23.3 (N = 69)
total	100 (N = 119)	100 (N = 66)	100 (N = 111)	100 (N = 296)

Thus, it is perhaps not wholly a surprise to find party activists divided over the question: just over half (51.4 per cent) favoured such restrictions, while the remainder was fairly equally split between those who were uncertain about them (25.3 per cent) and those who opposed them (23.3 per cent). Nearly a fifth (18.9 per cent) of those likely to vote against EU membership were also opposed to restrictions on free trade; this suggests that our fourth hypothesis, about the cause of free trade being a motive for some Social Democratic Eurosceptics, is not entirely without foundation. But this observation must be put into perspective. Whereas 42.9 per cent of probable Yes voters supported trade restrictions, as many as 57.7 per cent of probable No voters did so.

There is, then, some evidence that free trade was a reason for some Social Democratic activists to vote against Sweden's accession to the EU.

However, it was at best a marginal factor. Certainly, there is no evidence that supporters of EU membership in the party were any less committed to free trade; indeed, all the data suggest that they were more committed.

Hypothesis 5: greens v greys

As we have seen, the evidence that a materialist–post-materialist cleavage exists within SAP is reasonably strong on the observable evidence (for instance, the struggles over nuclear power and the Öresund bridge), though the hypothesised correlation between this and the party's division over Europe is based on more fragmentary evidence. A correlation was perceived by a fair proportion of the sample: 20.4 per cent considered the party's Eurosceptics to be primarily motivated by environmentalism. The five questions in this section sought to ascertain the extent of any cleavage between materialists, or "grey" Social Democrats, and non-materialists, or "green" Social Democrats. The responses are conveyed in table 4.6.

The first question asked simply, "If you had to choose, which do you think should take priority in Swedish policy?" In the responses, there did seem to be a very apparent split in the party on this fundamental issue. Just over half (51.4 per cent) thought that economic growth should be the priority; just under half (48.6 per cent) thought that it should be protection of the environment. Moreover, if the evidence of a connection between this fault line and the one over Europe is not overwhelming, it is certainly present and significant. Of those respondents inclined to vote Yes, 58.2 per cent had growth as their preferred policy priority, to just 41.8 per cent who preferred environmental protection. Of those likely to vote No, the distribution was reversed: 43 per cent preferred growth, to 57 per cent who preferred environmental protection. Similar results can be seen in the next question, which sought to get away from abstract concepts and to establish the existence of a materialist–non-materialist fault line on the basis of a substantive political issue: nuclear power. Nearly half of the respondents (48.5 per cent) were in favour of nuclear power in Sweden, nearly a fifth (19.8 per cent) were uncertain and nearly a third (31.7 per cent) were against. The division in the party on this crucial issue is thus clear. So too, if less emphatically, is the correlation between it and the EU question. Of probable Yes voters, 60.3 per cent were in favour of nuclear power, compared to just 41.1 per cent of probable No voters. Those against it, mean-

Table 4.6 Testing the green v grey fault-line

Referendum voting intention

	Yes %	Uncertain %	No %	total %

If you had to choose, which do you think should take priority in Swedish policy?

	Yes %	Uncertain %	No %	total %
Economic growth				
	58.2	53.7	43.0	51.4 (N = 146)
Protection of the environment				
	41.8	46.3	57.0	48.6 (N = 138)
total				
	100 (N = 110)	100 (N = 67)	100 (N = 107)	100 (N = 284)

Are you for or against nuclear power in Sweden?

	Yes %	Uncertain %	No %	total %
For	60.3	40.0	41.1	48.5 (N = 147)
Uncertain	18.2	32.9	13.4	19.8 (N = 60)
Against	21.5	27.1	45.5	31.7 (N = 96)
total	100 (N = 121)	100 (N = 70)	100 (N = 112)	100 (N = 303)

while, included 45.5 per cent of probable No voters, compared to only 21.1 per cent of probable Yes voters.

The hypothesis that suggested a correlation between, on one hand, the divide among Social Democrats over EU membership and, on the other, one between grey and green Social Democrats, has received some firm support in this survey. It is certainly the most "successful" of the hypotheses presented thus far.

Hypothesis 6: influencers v democrats

Our final hypothesis proposed an ideological fault line in SAP between those whose priority was the enhancement of Swedish national influence, for which EU membership offered more scope than any alternatives (including, for example, EEA membership), and those whose priority was the preservation of Swedish democracy, for which national sovereignty was considered indispensable. Four questions attempted to test this hypothesis (see table 4.7).

The sample was first asked to choose one of three statements relating to "your view of democracy". It did not succeed in highlighting much of a difference in priority among the respondents. Nearly nine out of ten (87.8 per cent) agreed that "it should not be compromised just for economic gain," with just over one in ten (12.2 per cent) thinking that "sometimes it is necessary to balance pure democracy with considerations of efficiency—for example, when managing the single European market." A distinction was observable when responses were controlled for voting intention; but their numbers were too small, when set against those who prioritised democracy over efficiency, to be very significant. Similarly, responses to the statement, "Democracy and national sovereignty are fundamentally interdependent," also failed to illuminate much of a divide. Over three-quarters of respondents (78.2 per cent) fully or partly agreed, with around a sixth (15.7 per cent) disagreeing. Again, there was some discrepancy between probable Yes voters (33.9 per cent of whom fully agreed, 23.7 per cent of whom disagreed) and probable No voters (51.8 per cent of whom fully agreed, 10 per cent of whom disagreed). But, once more, the general distribution of responses in support of democracy and national sovereignty's interdependence is more noteworthy.

Table 4.7 Testing the influence v democracy fault-line

Referendum voting intention

	Yes %	Uncertain %	No %	total %

Which, if any, of these statements most closely accords to your view of democracy?

It should not be compromised just for economic gain

	Yes %	Uncertain %	No %	total %
	81.2	87.9	94.6	87.8 (N = 259)

Sometimes it is necessary to balance pure democracy with considerations of efficiency—for example, when managing the European single market

	18.8	12.1	5.4	12.2 (N = 36)

"Democracy and national sovereignty are fundamentally interdependent."

	Yes %	Uncertain %	No %	total %
Fully agree	33.9	36.6	51.8	41.1(N = 123)
Partly agree	37.3	45.1	31.8	37.1 (N = 111)
Uncertain	5.1	7.0	6.4	6.0 (N = 18)
Disagree	23.7	11.3	10.0	15.7 (N = 47)
total	100 (N = 118)	100 (N = 71)	100 (N = 110)	100 (N = 299)

Which, if any, of the following statements most closely corresponds to your views on the EU and democracy?

The EU is a democratic organisation

	36.7	10.0	3.7	18.4 (N = 55)

The EU at present lacks democratic credibility, but it can be improved through reform

	63.3	68.6	33.9	53.8 (N = 161)

The EU is undemocratic and inherently incompatible with true democracy

	-	21.4	62.4	27.8 (N = 83)
total	100 (N = 120)	100 (N = 70)	100 (N = 109)	100 (N = 299)

"EU membership offers a means of reasserting political influence over multinational firms and international capital."

	Yes %	Uncertain %	No %	total %
Fully agree	25.4	6.9	0.9	11.8 (N = 36)
Partly agree	58.5	40.3	10.5	36.2 (N = 110)
Uncertain	7.6	22.2	12.3	12.8 (N = 39)
Disagree	8.5	30.6	76.3	39.1 (N = 119)
total	100 (N = 118)	100 (N = 72)	100 (N = 114)	100 (N = 304)

Finally, a question sought to test whether the sample considered that "EU membership offers a means of reasserting political influence over multinational firms and international capital." Nearly half (48 per cent) fully or partly agreed with the statement; not very many fewer (39.1 per cent) disagreed. While 25.4 per cent of probable Yes voters fully agreed and as many as 58.5 per cent partly agreed, the figures for probable No voters were only 0.9 per cent and 10.5 per cent respectively. While those disagreeing with the statement comprised fully 76.3 per cent of likely No voters, the proportion of likely Yes voters was 8.5 per cent.

It would see that the notion of a cleavage between Social Democrats who prioritised national influence, and those prioritised democracy, does find some supporting data. But it could be argued that no Social Democrat would disagree with the objective of attaining maximum democracy, or improving political influence over multinational firms and capital. What these questions identify are as much differences in judgment—about the scope for democratically reforming the EU, and its capacity for controlling the international economy—than of ideological objectives. Conclusions drawn in this section, then, must be especially tentative.

Conclusions

If there is one overriding impression from analysing the arguments deployed in the debate, it is that the stance of each side was multi-faceted. In as much as there can be "monolithic" ideological positions in any party, on any major political issue, the Social Democrats' discussion about the EU was its antithesis. There was at least some evidence to support all the six hypotheses offered in this chapter. Isolationist welfare nationalism did appear to motivate some opponents of membership. Some did think that the Union offered an inadequate basis for the radical-left Social Democracy that they espoused. Neutrality was very important to many likely No voters. A few activists did see membership as a violation of their commitment to free trade. And some did consider membership incompatible with democracy, either as the EU was currently constituted or because they made a fundamental connection between national sovereignty and democracy, or *folkstyre*. Yet none of these motivations was in isolation a convincing explanation of the divide within SAP. Indeed, aspects of its Euroscepticism were actually contradictory. An insistence on maintaining maximum free trade, for example, or support for EEA membership, can hardly be squared with fears for the EEA's effect on the Swedish welfare state, or a rejection of economic growth as the basic goal of public policy. In other words, the European debate in SAP was extremely eclectic, with unlikely coalitions forming around either support for or opposition to accession.

The data also suggest that there was little of the "European ideal" inspiring even many supporters of EU membership in SAP. Large numbers of probable Yes voters were ambivalent about a European identity; offered contradictory responses in stating their security-policy preferences; believed that accession might threaten Sweden's environmental policies; saw the EU as primarily a neo-liberal institution; and doubted whether the Union offered much of an opportunity for re-establishing political control over an internationalised economy. All this suggested that other, essentially negative reasons—perhaps most plausibly, the fear of isolation—may have persuaded them to fall behind their party leaders' position.

As far as our research questions are concerned, we may surmise that the prospect of EU membership did activate certain ideological cleavages, the roots and expressions of which can be identified by examining the history and development of the party. Of our six hypotheses, the two with most evidence to sustain them are probably those that concern left against right, and materialist against non-materialist. It is arguable, then, that a type of ideological coalition, comprising the traditionalist left and post-materialist Social Democrats, constituted the opposition to the party leadership's plans for Sweden to join the EU. The post-materialist element might be connected, in a way that the data presented here cannot illustrate, to SAP's long-standing support among the rural and small-town working class in Sweden. Alternatively, it could mean that something like Inglehart's generational theory of political values is in operation in the party[28]—except that, in contrast to what he proposed, it manifests itself in SAP as opposition to rather than support for European integration.

Yet the evidence for this Eurosceptical coalition presented in our survey is far from conclusive. It is certainly not sufficient to demonstrate that the existing ideological factions in SAP were the only, or even the main, cause of its division over EU membership. In any case, the question of why this issue, at this time, should cause these ideological divisions to be activated, when so many other political issues in the Social Democrats'

---

[28] Ronald Inglehart, *The Silent Revolution: Changing Values and Political Styles among Western Publics* (Princeton, NJ, Princeton University Press, 1967).

history had had no similarly stimulating effect, or a much less powerful one, remains unanswered.

An alternative method of explanation is thus presented in the following chapters. This and the previous chapter's method has been essentially inductive: it has observed the characteristics of the Social Democrats' divide and sought to posit an explanation for it on the basis of this observation. This approach has yielded valuable contextual information. We are more familiar now with the contours of the party's internal debate over Europe since the EC was formed, and in the 1990s especially. This knowledge is indispensable for any explanation of the motivation and calculation of the political actors involved. However, it is necessary now to consider a different approach. Using essentially deductive models of party behaviour, it may be possible to obtain a more satisfactory appreciation of precisely what lay behind the Social Democrats' difficulties with the issue of the European integration.

5   Crisis and volte-face

> Up to the crisis package in autumn 1990, membership was considered
> unthinkable by the Social Democratic leadership, with the EC seen as
> something like a symbol of capital and free-market forces. But after a
> couple of changes of direction in press statements, suddenly the official
> line was precisely the opposite. Sweden ought to seek membership. Today
> the EU is presented as the closest thing to a Social Democratic project that
> its own visionaries think can be realised.[1]

Having examined the history of the Social Democratic approach to European integration, an approach conditioned by years of political distance
from the Community and by the great success the party and the Swedish
economy had enjoyed, we can now appreciate the reasons why Social
Democrats might not have been naturally inclined to support the idea of
EC membership. Thus, it makes sense to turn around the question that we
have addressed in the preceding chapters. Instead of asking why significant
sections of the party opposed Swedish accession, we may instead ask why,
given the grass-roots' Euroscepticism, the leadership nevertheless changed
party policy in 1990 and announced that Sweden would be applying for
membership. The fact that its decision to do so caused dissent is actually
not surprising. The salient question is why, when so much of the party's
success had been founded on its internal unity, the leadership chose to risk
jeopardising that unity over Europe.

That the leadership's behaviour was closely connected to Sweden's contemporary economic difficulties is widely accepted. The nature
of that connection, however, is subject to much more disagreement, with at
least three different analyses proffered. One is that due to the development
of the EC—an exogenous factor—Sweden could no longer afford to stay
outside it. According to this view, Sweden's economic problems were in
large part attributable to non-membership, and the consequences of that
situation continuing were even more unattractive to the Social Democratic
leadership than the danger to party unity that a membership application
would entail.

---

[1] *Veckans Affärer* June 13th 1994.

A second explanation concerns Social Democratic political economy, and how its intellectual and practical efficacy had become exhausted amid a changing international economy. Application to the EC thus constituted a commitment to a new economic strategy. In his book on Swedish parties' historical and contemporary views on European integration, Gidlund reflects on the rapid change in the country's policy that was enacted in 1990.

> What in the end made the Social Democratic government decide to announce an application for EC membership will surely be the subject of several studies and much research. Here I content myself with formulating the hypothesis that the economic crisis was a catalyst for the decision. The Social Democratic government's difficulty in finding new solutions to the economic problems of the day had become much clearer. The old responses, involving state support, and therefore increased public spending, were no longer thought to work in an economy and labour market adapted to and dependent on the rest of the world.[2]

This explanation is not wholly incompatible with the first one, but nevertheless places its emphasis on an endogenous factor: the failure of Social Democratic economic management. A third explanation is, in its turn, reconcilable with this argument; but it gives most weight to short-term political circumstances. In short, it suggests that the economic crisis that Sweden suffered at the end of the 1980s and the beginning of the 1990s created extremely acute political difficulties for the Social Democratic leadership, and that its rapid change of European policy was prompted by a measure of political desperation. Any change of European policy, however cautious and gradual, was likely to have risked discord within the party. In the event, such a change of position was completed, even at a generous estimation, within perhaps three months. The decision to apply to the Community was more a reflection of this desperation than a sign of the intrinsic value of the new policy *per se*.

The third explanation is most favoured here. In order to justify this view, the remainder of the chapter is divided into two parts. The first is an analysis of the nature of the crisis of the Swedish economy. If it was caused mainly by factors exogenous to Sweden, a change in the country's external position—that is, EC membership—might have been necessary to deal with it, thus explaining the behaviour of the Social Democratic leader-

---

2 Gidlund, *Partiernas Europa* (1992), pp.56-57.

ship. If, on the other hand, the crisis had been caused mostly by endogenous factors, the implication would be that the answer did not necessarily lie in joining the EC. Such a conclusion would not only give credence to the second and third explanations outlined above, it might also help to explain why SAP's naturally Eurosceptical temperament was and remains resistant to change. The second part of the chapter applies some of the theories of party behaviour described in chapter 2 to the detailed circumstances surrounding the change of policy. This should help us to gain greater appreciation of how and why policy changed so quickly—and thus why the party leadership failed to carry the rank-and-file.

What type of crisis?

In the view of some observers, Sweden's path to the Union became clear, albeit retrospectively, soon after it announced in 1990 its intention to apply for membership. Just as Europe's changed geopolitics cast a new light on Swedish security policy, it was argued that something similar had happened in the field of economic policy. The flood of capital out of the Swedish economy in 1985-90, much of it to the EC, gave the Social Democratic government "little option" but to apply for membership,[3] due to a "(belated) recognition that feasible alternative strategies are simply not available".[4] Further support for this "inevitablist" view is inferred from the way foreign direct investment in the country recovered markedly in 1991, once the application for EC membership had been announced.[5] However, it is possible to see a rather less direct relationship between, on one hand, geopolitical and economic change, and, on the other, the European policy of Sweden's Social Democrats.

The end of the cold war did not create a Swedish interest in joining the EC of the type that applied elsewhere in Europe. Sweden had long reached an understanding with NATO that the alliance would help it in the

---

[3] Jonas Pontusson, *The Limits of Social Democracy: Investment Politics in Sweden* (Ithaca, NY, Cornell University Press, 1992), p.119.
[4] John F.L. Ross, "Sweden, the European Community, and the Politics of Economic Realism", *Cooperation and Conflict* vol. 26, no.3, 1991, p.125.
[5] Stråth (1992), pp.102-7.

case of a Soviet attack,[6] so the fall of communism did not present the type of chance that it did for Austria and Finland. For these countries, an application to the EC represented a (possibly fleeting) window of opportunity to reaffirm western ties and identity, without being overtly provocative to Moscow.[7] The end of the cold war can be seen as a *facilitator* of Swedish EU membership, but not an active *incentive* to seek it.

As far as economic interest is concerned, access to the continental market was certainly vital to Swedish exporters. But by the time of the EU referendum the danger of marginalisation from the EU market had arguably been averted. The EEA came into force in 1994 and, months before the referendum, Ingvar Carlsson agreed: "If we are really honest, we have certainly resolved the economic part [of Sweden's position] through the EEA agreement."[8] Moreover, in April 1996 it emerged that the Swedish government had succeeded in persuading the GATT's secretariat to delay until a day after the EU referendum publication of a report in which it warned that Sweden's adoption of the EU's common external tariff would actually raise the cost of some imported consumer goods, as Swedish tariffs were lower. Textile imports, for example, which Sweden had deregulated in 1991, became subject to new EU tariffs and quotas.[9]

In any case, the EU amounts to far more than the single market, and several obligations of membership were far from advantageous to Sweden. The EEA's political limitations notwithstanding, in it Sweden escaped some of the EU's least favourable aspects. One such was incorporation into the CAP, which required a large adjustment to—indeed, in some respects, a reversal of—reforms enacted in Sweden after 1990. These had been designed to deregulate Swedish agriculture so as to have production reflect demand more closely, but the CAP forced Swedish farms to operate in a still more controlled market. A commission appointed by the non-socialist coalition in 1991 to diagnose the causes of Sweden's economic problems, which had the economist Assar Lindbeck as its chair and

---

6 Andrén (1991), p.79. Recent investigations not only confirm this implicit American guarantee of Sweden's security (*Svenska Dagbladet* August 3rd 1998), but further suggest that the Soviet Union was well aware of it (*Svenska Dagbladet* August 7th 1998).

7 René Schwok, "EC-EFTA Relations, A Critical Assessment", paper presented to the Second Pan-European Conference in International Relations, Paris, September 1995, p.9; David Arter, "The EU Referendum in Finland on 16 October 1994: A Vote for the West, not for Maastricht", *Journal of Common Market Studies* vol. 33, no. 3, 1995.

8 *Veckans Affärer* June 13th 1994.

9 *Dagens Nyheter* April 4th 1996.

which included two non-Swedes, pointed to agriculture as "an area in which future EC membership will entail a less desirable policy for consumers than what we [Sweden] might achieve on our own. Sweden has started to deregulate its agricultural sector. EC membership will mean re-regulating it."[10]

In addition, Sweden immediately became a net contributor to the Union's budget, and this deficit rose further in 1999 when a derogation from paying its full contribution expired. Thereafter Sweden's annual budget contribution was forecast to be around SKr20 billion, the equivalent of 4 per cent of central-government spending. Only half that figure was expected to flow back into the country from EU coffers, mainly through farming and regional subsidies. In the context of fiscal crisis and retrenchment of public expenditure, paying a sizeable EU "membership fee" was not obviously attractive to many Social Democrats.

Finally, the observation that Sweden's poor rates of economic growth reinforce the notion of an inexorable movement towards EU membership is also inadequate to demonstrate a causal relationship. Various economists criticised the conclusions of another commission appointed by the non-socialist government, which reported in January 1994 on the economic consequences of EU membership. The commission suggested that secure access to the EU's internal market would encourage extra investment in the domestic economy to the tune of 0.9-1.2 per cent of GNP, and that it was "not unreasonable" to suggest that this would boost annual growth by 0.5 per cent. These projections were questioned mainly on the grounds that "we simply do not have enough information—theoretical and empirical—to quantify all the effects."[11] The Lindbeck commission, while seeing benefits in the greater economic competition likely to be promoted by EEA membership, remained neutral on whether full EC membership would promote or impede the reforms it recommended.[12]

Conversely, the argument that Sweden's economic crisis had largely domestic rather than European or global causes, although only briefly outlined here, is also plausible. Growth in Sweden had been slug-

---

10 Assar Lindbeck, Per Molander, Torsten Persson, Olof Petersson, Agnar Sandmo, Birgitta Swedenborg and Niels Thygesen, *Turning Sweden Around* (Cambridge, Mass., MIT Press, 1994), p.87.

11 Ari Kokko, "Sverige: EU-medlemskapets effekter på investeringar och tillväxt", in Magnus Blomström and Robert E. Lipsey (eds), *Norden i EU: vad säger ekonomerna om effekterna?* (Stockholm, SNS förlag, 1994), p.28.

12 Lindbeck *et al* (1994), pp.214-15.

gish since the early 1970s, suggesting that the economic crisis that came to a head in the 1990s had been building for a long time, and that the causes were essentially home-grown. Between 1952 and 1973 growth had averaged 3.9 per cent; between 1973 and 1990 it was just 1.6 per cent, markedly lower than the average in the rest of the OECD.[13] Partly in consequence, but also because the size of the public sector had made the economy especially vulnerable to the external shocks of the 1970s, public deficits and state debt escalated throughout the 1980s. In short, Sweden suffered from gathering crises of competitiveness, public finance and the balance of payments.

It is also clear that the economic strategy pursued by the Social Democrats after they returned to power in 1982—yet another "third way" (this time between the British and French policy responses to recession)—had, after a hopeful start, gone badly wrong. It aimed to create an export-led recovery, achieved, first, through an aggressive "super-devaluation" of the krona by 16 per cent. The strategy required workers to forgo wage compensation for the devaluation, and thus accept a cut in real wages to make up the competitive disadvantage. At first, the trade unions appeared to deliver this. But centralised wage-bargaining had broken down and, particularly with corporate profits soaring, the means of preventing individual unions from seeking nominal wage rises became increasingly ineffective. Inflation was high in Sweden throughout the 1980s, which soon eroded gains in competitiveness. In addition, it is hard to contradict the assertion that liberalisation of the Swedish credit markets in 1985 was disastrously timed, as it aggravated an inflationary explosion in domestic consumption.[14]

As for the basic causes of the malaise, emphasis is commonly placed on the erosion of wage discipline, but interpretations of why this happened vary. Monetary policy is often thought to have been too loose from the early 1970s, which allowed inflation to undermine competitiveness; but the importance of fiscal policy, particularly the size of the tax burden, is subject to more dispute. Another factor is sometimes seen in the Swedish labour movement's becoming radicalised before and during

[13] Lars Pettersson, "Sweden", in David A. Dyker (ed.), *The National Economies of Europe* (London, Longman, 1992), p.162.

[14] Andrew Martin, "Macroeconomic Policy, Politics, and the Demise of Central Wage Negotiations in Sweden", Centre for European Studies Working Paper Series 63 (Cambridge, Mass., Harvard University, 1996), pp.16-26.

Palme's leadership of SAP. It may have overreached itself politically in the 1960s and 1970s by using legislation to interfere with firms' management, which, it is suggested, provoked the employers into withdrawing their co-operation in wage-formation.[15] Then there are microeconomic factors. Some point to long-term changes in production, which required more flexible wage-formation at a lower level, and at markets becoming more internationalised, which encouraged Swedish firms to move production to countries that were nearer customers and that were friendlier to profits.[16]

Alternatively, an important critique of the whole Swedish model in the mid-1980s attributed cost-push inflation largely to the difficulty of sharing out a pie that was diminishing in size. This predicament was itself attributable to the accumulated, growth-retarding effects of distorted price signals, not least by the large public sector and by wage solidarity.[17] Finally, there seems little doubt that the abolition of controls on capital movements in 1989 was crucial, if only in amplifying the effects of deeper-seated problems. It seems that controls were being increasingly circumvented anyway, but liberalisation greatly expanded the scope for Swedish firms to exploit a new option, that of "exit". If they did not like the economic climate in Sweden they could freely relocate abroad—which, judging by the immediate outflow of capital, they did.

Yet what these short- and long-term problems have in common is that all were either essentially domestic in character, or had no obvious means of redress in the EC—*pace* the argument that exclusion from the Community was in itself a cause of Sweden's economic problems. Indeed, the Social Democratic government itself made it plain in 1989-90 that it viewed the country's major economic problem as wage inflation, an endogenous phenomenon.[18] The Community would not mitigate changes in

---

15 Klas Åmark, "Social Democracy and the Trade Union Movement: Solidarity and the Politics of Self-Interest", in Klaus Misgeld, Karl Molin and Klas Åmark (eds), *Creating Social Democracy: A Century of the Social Democratic Labor Party in Sweden* (Pennsylvania, Pennsylvania Universiy Press, 1992), p.92.

16 Stråth (1992), p.:59ff; cf. Pontusson (1994), pp.42-51; Magnus Ryner, "Economic Policy in the 1980s: The 'Third Way'. the Swedish Model and the Transistion from Fordism to Post-Fordism", in Wallace Clement and Rianne Mahon (eds), *Swedish Social Democracy: A Model in Transition* (Toronto, Canadian Scholars' Press, 1994).

17 Erik Lundberg, "The Rise and Fall of the Swedish Model", *Journal of Economic Literature* vol. 23, 1985, pp.28-32. See also Jan-Erik Lane and Svante Ersson, *Comparative Political Economy: A Developmental Approach*, 2nd ed. (London, Pinter, 1997), pp.205-13.

18 Per Löwdwin, *Det dukade bordet. Om partierna och de ekonomiska kriserna* (Uppsala, Acta Universitatis Upsaliensis, 1998), p.297.

the nature of capitalist production (even if they were understood by policy-makers); nor would it rein back the distortions flowing from an oversized public sector (if that was indeed a real problem); nor would it prevent firms and capital emigrating in search of better returns.

Of course, none of this is to suggest that there was no Swedish interest in joining the EC, even if membership did have its disadvantages, and even if Sweden's basic economic problems were probably not attributable to its non-membership. It is arguable that non-tariff barriers to Swedish exports were a potential danger, especially before the EEA was agreed. And the argument that if it was to be subject to them, Sweden should have a say in deciding the rules of the EC market is clearly a strong one. Moreover, as we shall see in greater detail in chapter 7, Community membership might have offered a new lodestar for Swedish monetary policy. Yet even this, it could be argued, was but one possible method of addressing Sweden's problem of chronically high inflation; there were other strategies available that did not depend on the EC, or even on any external discipline. The point of this section is merely to suggest that the structural geopolitical and economic factors militating in favour of Swedish accession were, objectively speaking, not so suddenly overwhelming or urgent as to explain the Social Democratic leadership's behaviour in the question during 1990.

There is no doubt that whatever the cause of the Swedish economic crisis, the country's big export firms were very keen on EC membership, and that they exerted pressure on the government to pursue the option. Their concentration, organisation and traditionally close ties to the Social Democratic elite gave them the means to do this. This is the basis of Ingebritsen's explanation of why the Swedish government changed its European policy.[19] However, our focus is specifically on the Social Democratic Party. Its leadership was subject to lobbying from industry, but the grass-roots would naturally have been less persuaded by the blandishments of big business—especially if, as is arguable, the Swedish economic crisis was largely domestic in origin, and thus was not obviously soluble by anything the EC could offer. A Eurosceptical Social Democratic economist later made exactly this point: "[W]hen domestic consumption grew in 1986-90 it created an inflationary economy *in Sweden*, and the reduction in domestic consumption created a deflationary economy *in Sweden*." This,

---

[19] Christine Ingebritsen, *The Nordic States and European Unity* (Ithaca, NY, Cornell University Press, 1998).

he added, "effectively refutes the theory that national Keynesianism, that is, national demand-management, no longer gives results".[20]

An adapted model of two simultaneous arenas

If we look, then, to the party itself for the primary causes of the precipitous change in the leadership's European policy, theories of party behaviour can assist our analysis. The adapted model used here envisages two simultaneous arenas in which the Social Democratic leadership was playing during the events of 1990. Constructing conceptual models is a question of balancing the requirements of parsimony, which requires simplicity in the cause of analytical rigour, and realism. In this case, emphasis can shift towards parsimony.

Of the three goals ascribed to party leaders by Harmel and Janda (see chapter 2), the one concerning implementation of policy can also be disregarded when the issue of Swedish EC membership is considered. This is because SAP arguably had no European policy that was fundamental to its political programme. The Liberal and Moderate parties were keenest on closer relations with the Community; but up to 1990 no parliamentary party, socialist or bourgeois, had gone so far as to campaign for Sweden's accession. And while the Centre and Social Democratic parties were undoubtedly more sceptical, they had not tied themselves programmatically to opposing membership. Such was the fluidity of the situation that, conceivably, any of those parties could have adopted the position of advocating membership. At least, none could feel debarred from taking it just because their rivals supported it.

Thus, two simultaneous and (in Tsebelis's terms) nested games envisaged by our model are: (a) an *intra-party* game, in which the party leadership interacts with its membership and with trade unions; and (b) an *inter-party* game, in which the leadership interacts with leaders of other parties, all of whom compete for the support of the electorate. But two important nuances are appended to the second game. First, it is assumed that the party in question is in government. Second, the indirect but crucial importance of the financial markets is recognised formally. (See figure 5.1.)

---

20 Wibe (1994), p.203, emphasis in original.

Figure 5.1 Simplified model of leadership's location in determining social democratic party strategy

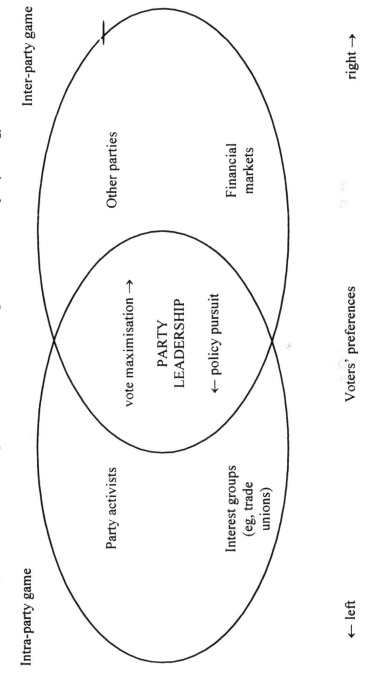

Intra-party game

Inter-party game

Party activists

Other parties

Interest groups
(eg, trade
unions)

Financial
markets

vote maximisation →

PARTY
LEADERSHIP

← policy pursuit

← left

Voters' preferences

right →

The reason why these nuances have been added is that, since the end of the 1980s especially, a government's interaction with international financial markets has become too important to ignore.[21] Forced devaluation—unlike that implemented by an incoming Social Democratic government in Sweden in 1982, which was part of a deliberate, aggressive economic strategy—necessarily signals a policy defeat. Moreover, there are two other reasons why the Swedish government at the beginning of the 1990s may have been particularly vulnerable in political terms to a forced devaluation of the krona. First, monetary policy had been given a significantly changed role since the middle of the previous decade. Rather than being mainly a tool of counter-cyclical macroeconomic policy, its primary aim became to keep inflation down, as the Social Democratic government implicitly gave up relying on the labour market parties to co-operate in keeping inflation in check.[22] Given that the 1982 devaluation had led to severe overheating in the Swedish economy,[23] further devaluation would probably also have had inflationary consequences, thus reducing competitiveness further, and also forced interest rates higher to contain inflation, thus dampening growth and putting pressure on employment. Second, by the mid-1980s budget deficits had accumulated to produce a public debt sufficiently large, and with a sufficient proportion of it held by foreign creditors, to make devaluation an even more expensive prospect. Either outcome would have been disastrous, economically and therefore politically, for the government.

This is why our model explicitly recognises the important role of the international financial markets in the inter-party arena, in which the Social Democratic Party leadership interacted with other parties' leaderships in attempts to maximise electoral support. The dependent variable is *the leadership's strategy*: in seeking voters' support, which does it prioritise, the intra-party game or the inter-party game? The hypothesis is that *its priority changed as environmental factors changed*. The combination of short-term, external political events and a shrewd campaign by the opposi-

---

[21] As Hinnfors and Pierre acknowledge in their analysis of a later Swedish government's abortive battle to maintain the krona's exchange rate, "The awkwardness for the government of designing its policies for a group of actors which does not have any identifiable (and indeed, which can normally not even be identified in the first place) political goals is amazing." Jonas Hinnfors and Jon Pierre, "Currency Crises in Sweden: Policy Choice in a Globalised Economy", *West European Politics* vol. 21, no. 3, 1998, p.116.

[22] Lundberg (1985), p.27; Ryner (1994), p.265.

[23] Lindbeck *et al* (1994), p.30; Martin (1996), pp.9-14.

tion parties contributed significantly to a domestic climate in which the government was consistently forced into rushed, ad hoc policy-making.

A gathering economic crisis

The "war of the roses"

By the end of the 1980s it was clear that the economy was seriously overheating. This in turn provoked bitter recrimination, a "war of the roses", between SAP's leadership and the trade unions, the latter being particularly opposed to the government's plans to put 2 per cent on VAT in the hope of dampening consumption. In the face of this opposition, the government was forced to forge a pact with the Centre Party in which it was agreed that, from September 1989 to the end of 1990, 3 per cent of all wages would be "obligatorily saved". LO and elements in SAP, as well as the Liberals and Moderates, were strongly against the deal. In January 1990 Kjell-Olof Feldt, the Social Democratic finance minister, warned that the economy had reached crisis point, as inflation approached double figures, the current-account surplus dwindled and a SKr50 billion budget deficit was forecast for 1991. Although the government agreed with the Liberals the so-called "tax reform of the century", which cut marginal income-tax rates by shifting the tax burden onto capital and a broadened VAT base, negotiations between government, business and trade-union organisations—which Feldt referred to as the "last chance for the Swedish model"—broke down in early February.

An emergency austerity package proposed by the government a week later seemed to have won LO leaders' support. But when the chair of the Municipal Workers' Union, Lillemor Arvidsson, learned that as well as imposing a freeze on rents, dividends, local taxes, prices and wages, the "stop package" also proposed, remarkably, the temporary banning of strikes from February 15th, with fines for transgressors, she refused to go along with the agreement. (Her union had planned a strike for February 14th.) Suddenly, LO's leaders seemed to realise how consenting to a ban on strikes would be perceived by their rank-and-file. Without LO's support for the package, and with the Communist Left and the Moderates joining an unlikely parliamentary coalition against it, Carlsson resigned as prime minister; Feldt retired from politics. Within a week, however, Carlsson had formed a new government. The ban on strikes and the wage freeze were dropped, but the episode was a disaster for the labour movement in general, and SAP in particular. In its aftermath around 2,000 local-government employees were reported to have left the party. Local Social Democratic parties from all over Sweden threatened to withhold their contributions to its national coffers.[24]

In April the government agreed with the Liberals another austerity pact, which included a rise in VAT of 1 per cent and a cut in employers' social-security charges. The trade unions were unhappy enough about that.[25] But it was the shelving of plans for extended holiday-entitlement and parental insurance that was most politically sensitive. Anna-Greta Leijon, a member of the Social Democratic Executive Committee, resigned as chair of the Riksdag's Finance Committee over this breaking of her party's election promises. Meanwhile, while some in the party—in particular, the Association of Christian Social Democrats (the "Brotherhood Movement"), the Women's Federation and SSU—remained strongly committed to phasing out nuclear power, others were beginning to sound wary of the economic consequences. Many in SAP were also very dubious about the leadership's support for building the bridge over the Öresund to Denmark. All this was against the backdrop of rising controversy about the numbers of asylum seekers coming to Sweden, and a spate of racist attacks on them. Labour relations, meanwhile, had reached a very low ebb, especially in the public sector, which brought the government directly into conflict with important trade unions. Teachers, for example, took strike action in pursuit of a double-figures pay claim. The labour movement's traditional Mayday celebrations were marred by angry trade-union demonstrations against the Social Democratic government.

With this crisis mounting, Social Democratic leaders' attention was fixed firmly on the intra-party arena. Certainly, the party's popularity among voters (that is, its performance in the inter-party arena) was plummeting. A Temo poll undertaken before the stop package was presented showed SAP's support at just 32 per cent, down 6.2 per cent on the previous month. By the end of March a Sifo poll showed a similar figures—the lowest figure for the party ever registered by that polling firm. Two months later the state statistical bureau, Statistics Sweden, estimated that SAP had lost the support of 600,000 voters since the election in September

---

24 *Nordisk Kontakt* 5/90, April 1990.
25 *LO-tidningen* April 16th 1990.

1988. A newly formed left-wing party, the Workers' List, was preparing to put up candidates in the 1991 election, and openly targeted disillusioned Social Democrats. But the source of SAP's weakness at that time was its internal disunity, and its bitter dispute with LO. It was this that commanded the leadership's attention. Thus, in early 1990 a change of EC policy was not a priority for the party leadership. It would have been far too provocative to the party's activists, who were already unhappy with their leaders' handling of other issues.

But *two environmental changes*, relating to the inter-party arena, forced the party leadership to reassess its calculation. First, there was a highly effective campaign waged by the Moderates and Liberals, which caught the drift of pubic opinion and posed a considerable threat to SAP's electoral prospects. Second, and perhaps most importantly, there was a growing crisis of market confidence in the Swedish government. Both these environmental changes forced the Social Democratic leadership to reverse its priority, and turn its attention to the inter-party arena.

The emergence of the EC issue

There had been discussions about Sweden's relationship with the EC at all levels of the labour movement. LO, for example, had for some years had a committee devoted to European matters, and as early as spring 1987 it had been reactivated by the appointment to it of Gudmund Larsson, who had long been sympathetic to the Community. He had been resigned to the fact, however, that foreign-policy requirements precluded the option of full membership, and acknowledged that European policy posed difficult questions for Swedish Social Democrats.[26] When, in spring 1989, the LO committee prepared to present its analysis of the situation to the confederation's national executive, Larsson's pro-EC inclination was very much a minority view, and he did not push the issue.

The fall of the Berlin wall in November 1989 naturally changed the situation. In spring 1990 Larsson and the chief economist of the Metal-Workers' Union, Håkan Arnelid, drew up a further list of European policy

---

[26] "In approaching the EC, we risk losing our special character," he admitted in early 1989. "Sweden has a relatively equal division of income, a relatively equal wage-structure and a progressive tax system. The policy of distribution seems different on the continent." *Aktuellt i Politiken* February 23rd 1989.

options for the Swedish labour movement. At that time, some leading figures in SAP were beginning tentatively to broach the possibility of closer
relations with the EC. Odd Engström, the deputy prime minister, told the
LO newspaper:

> Personally, I think that it would be fantastic if Swedish Social Democracy
> can take part in building a united European labour movement...The East
> European states are very interested in the European Community. We must
> be careful that Sweden is not the last to take down its guard when it
> comes to defence policy.[27]

But by the spring the Social Democrats, beset by so much discord both
within the party and with the trade unions, had no desire to see activated
another issue that clearly had the potential to cause internal controversy.
When, in March, the Danish foreign minister, Uffe Elleman-Jensen, had
suggested joint applications to the EC by the three Nordic members of
EFTA, his Swedish counterpart, Sten Andersson, reacted coolly, even
sharply. Andersson's opposition to any such move became apparent to his
colleagues on the Social Democratic Executive Committee in May and
June 1990, when it was discussing the motions that were to be debated at
the party's forthcoming congress in September.

At the end of April Carlsson was criticised at a meeting of the
Nordic labour-movements' co-operation committee (Samak) in Helsinki
by the Icelandic foreign minister for not using the Swedish presidency of
EFTA to push the EEA negotiations with sufficient enthusiasm. Carlsson
rebutted what he called this "incomprehensible" criticism, and insisted,
"We want to be part of all the four freedoms [that the EC's single-market
programme involved]." But he also reaffirmed that co-operation with the
Community in the fields of foreign and defence policy was ruled out.[28]
And when, at the publication of his and Arnelid's memorandum to LO's
executive, Gudmund Larsson spoke openly about the possibility of Sweden's eventually joining the EC, Rune Molin, the industry minister, afterwards berated him for what Molin considered an indiscretion.[29]

Then, quite suddenly, the European issue became the biggest in
Swedish politics, with an intense debate being conducted by leading politi-

---

[27] *LO-tidningen* May 19th 1990.
[28] *LO-tidningen* May 11th 1990.
[29] Stated by Larsson in an interview with the author, June 1997.

cians, in large part through debate articles written in Sweden's biggest-selling broadsheet newspaper.

On May 27th Carlsson seemed firmly to suggest that his government was not inclined to pursue the membership option. In a *Dagens Nyheter* article that ran under the headline "'EG-medlemskap omöjliggörs'" (EC membership made impossible), he argued that neutrality was still indispensable for Sweden's security. Three weeks previously, at the EC's Dublin European Council, closer foreign policy co-operation had been stated as an objective. "Should the EC really move towards such far-reaching co-ordination," the prime minister maintained in his article, "it would not be possible for Sweden to pursue the question of membership." The country was, he argued, actively involved in European co-operation in the fields of trade, through the EEA negotiations, and security, through the Conference on Security and Co-operation in Europe (CSCE), which gave it "the power to be as good a European as all the others".30

If Carlsson's intervention had been designed to dampen political speculation about the possibility of EC membership, it had precisely the opposite effect. The next day Bengt Westerberg, the Liberal leader, urged a Swedish application during the next parliamentary term. Importantly, international opinion seemed to back him up. Elleman-Jensen, attending a liberal conference in Sweden, mentioned the potential of 16 Nordic votes in the EC's Council of Ministers, and played down the idea of Swedish neutrality being an obstacle to membership. A Community representative at the conference, Enriqué Baron Crespo, promised a "sympathetic reaction" to any Swedish application. Two days later a Liberal economist and MP used *Dagens Nyheter* to make a strong personal attack on the prime minister, accusing Carlsson of setting conditions for joining the EC "so strict that in practice one must now rule out the possibility that Swedish Social Democracy will work for membership", and of elevating neutrality to an end in itself, rather than as a means towards peace in Europe. This, he claimed, was "a deeply immoral policy".31

Meanwhile, events beyond Sweden's borders were contributing still more to the political momentum that was taking the country closer to an application to the EC. On May 30th Boris Yelstin became Russia's first elected president. On June 1st Presidents Bush and Gorbachev initialled the START arms-reduction agreement and a week later the states of the

---

30 Ingvar Carlsson, "'EG-medlemskap omöjliggörs'", *Dagens Nyheter* May 27th 1990.
31 Carl B. Hamilton, "'En omoralisk Europapolitik'", *Dagens Nyheter* May 31st 1990.

Warsaw Pact agreed to wind up their military co-operation. On June 12th Russia declared itself a sovereign state. On June 15th the Soviet foreign minister, Eduard Shevardnadze, wrote to 13 European states, including Sweden, proposing a nuclear-free Baltic region. On June 17th, Lithuania announced its intention to suspend its unilateral declaration of independence, which prompted the Soviet government to lift its blockade of the republic, and helped to reduce tension in that region.

The same day Carlsson attended a meeting of Nordic heads of government in Gothenburg, at which the Danish prime minister in particular spoke encouragingly about Nordic applications to the EC. A Soviet political scientist also wrote a newspaper article in which he suggested that Moscow, far from opposing Swedish accession to the Community, might even welcome it, as "it could contribute to greater mutual understanding between the Soviet Union and the EC."[32] Over the next three months, the superpowers agreed further arms cuts, as well as the terms of German reunification (which duly occurred on October 3rd). On July 1st German monetary union took place. All this had a predictable effect on public opinion. "The political climate was becoming more right wing in Sweden", Pedersen recalls, "and everything 'European' was becoming fashionable in intellectual circles."[33] By June the EC issue was burning too brightly— thanks largely to the efforts of the Moderates and Liberals—for other matters to displace it from the top of the political agenda.

Westerberg, the Liberal leader, went so far as to suggest a three- to five-year plan for Sweden's accession to the EC, and dismissed as "pathetic"[34] the caution on the issue displayed by the Centre Party leader, Olof Johansson. Johansson had furiously attacked the Liberals and Moderates for jeopardising both Sweden's neutrality and its negotiating position in the EEA talks. He had asked of Westerberg and Bildt, "Do you feel no responsibility for Sweden as a nation?"[35] But even the Centre leader soon began to change his tone. From her analysis of parliamentary debates in Scandinavia, Kite concludes that between 1987 and 1990 the Centre had

---

[32] Konstanti Voronov, "Sovjet välkommer svensk EG-anslutning", *Dagens Nyheter* June 30th.

[33] Thomas Pedersen, "EC-EFTA Relations: An Historical Outline", in Helen Wallace (ed.), *The Wider Western Europe: Reshaping the EC/EFTA Relationship* (London, Pinter, 1991), p.87.

[34] Bengt Westerberg, "'Patetiskt, Olof Johansson!'", *Dagens Nyheter* June 12th 1990.

[35] Olof Johansson, Pär Granstedt and Per-Ola Eriksson, "'Ni tar inte ansvar för Sverige'", *Dagens Nyheter* June 10th 1990.

been the most Eurosceptical of all the Swedish parties bar the Communist Left.[36] Yet a sign of the way the political wind was blowing was the decision by the Centre's National Executive on June 19th 1990, after a long debate, to keep open the party's position on EC membership. By late June even Johansson, undoubtedly keen to maintain at least a degree of unity among the non-socialist parties was accusing the media of portraying the Centre as anti-EC.

Reassessing priorities

It was the emergence of the European issue—due to a considerable degree to short-term events outside the country—that gave the Moderates and Liberals the means to press home their advantage most effectively. Both parties had long been the most enthusiastic advocates of closer Swedish ties to the Community, and intensifying their advocacy could be done without it smacking of political opportunism. But what made the issue so potent for them was the fact that their pro-EC line accorded with the trend in Swedish public opinion. Polls showed numbers supporting accession rising to 60-70 per cent—a remarkable figure in the light of the levels of public opposition to membership recorded after mid-1991.[37]

Faced with this pressure in the inter-party arena, in which external events were interacting with and being exploited by the campaigning of the Moderates and Liberals to win voters' support away from the Social Democrats, leading figures in SAP began to sound more open to the Community. An early pioneer was a former under-secretary of state, Sverker Åström, who was closely associated with SAP, although not formally a member. He wondered whether, Carlsson's article notwithstanding, neutrality might actually be compatible with EC membership.[38] He did this after consultation with leading members of the government—although not with Andersson, the foreign minister. Carlsson himself then insisted that "those who declare that I shut the door to EC membership are mistaken."[39] On June 7th the prime minister even cautiously welcomed a call by West

---

[36] Kite (1996), p.128.
[37] Jan O. Berg, "European Union Relations in Swedish Public Opinion: An Overview", in Centre for European Policy Studies, *The Fourth Enlargement: Public Opinion on Membership in the Nordic Candidate Countries* (Brussels, CEPS, 1994). pp.87ff.
[38] Sverker Åström, "'En onödig inskränkning'", *Dagens Nyheter* June 1st 1990.
[39] *Nordisk Kontakt* 8-9/90, July 1990.

Germany's Chancellor Kohl for a "United States of Europe", saying that Sweden might join if the threat of war in Europe had disappeared. On June 12th the Riksdag debated the EC question, after which *Dagens Nyheter* observed: "The government's attitude is changing, and contributions from LO and TCO people suggest that even those who have been reserved towards the EC are undergoing a change of mind."[40] Even then, however, Carlsson declared that, at that time, "it is not sensible...for Sweden to seek EC membership."[41]

During this period, with great pressure on Social Democratic leaders in the inter-party arena, there was some dispute among them about the correct response. Some clearly wanted to stick to the party's sceptical position, and at that time the foreign secretary was their most influential representative. At the start of June Andersson asserted that "if the EC becomes a union with total co-ordination of foreign and security policy, then it will be quite impossible for Sweden to become a member."[42] Others in the party's Executive Committee, however, were coming round to a different view. Bo Toreson, the party secretary, had for some time been open to the idea of exploring the possibility of Sweden's joining the Community (although, despite encouragement from Gunnar Stenarv, the party's international secretary, amongst others, he had not pressed the issue previously in the party's National Executive or Executive Committee). Just as importantly, Stig Malm, who had caused so many problems for the party in his capacity as chair of LO, was also keen on a new approach.

But what afterwards became seen as a decisive signal in the government's volte-face came on July 5th, when Carlsson published another article in *Dagens Nyheter*, under the headline "'EG hinder kan undanröjas'" (EC obstacles can be removed). It restated difficulties inherent in Europe's geopolitical situation and the level of the EC's foreign and security policy co-ordination, as he had seen them five weeks previously. Yet the prospect of these problems being surmounted was addressed less in terms of a hypothetical situation, and more as a realistic scenario. Carlsson repeated his government's formal position, as defined by the Riksdag, that "Swedish membership of the Community is not a goal of the ongoing discussion with the EC." But he went on:

---

40 *Dagens Nyheter*, "Sverige närmare EG", June 12th 1990.
41 *Dagens Nyheter* June 16th 1990.
42 *Dagens Nyheter* June 2nd 1990.

This applies *today*. As for the *future*, I have said—both in the Foreign Affairs Council and publicly—that Sweden might in certain circumstances seek membership of the EC. If the motivation for our neutrality should disappear, in that a thoroughgoing change in the strategic military situation removes the risk of war in Europe—when the lion lies down with the lamb, so to speak—then one would be free to do things that are not conceivable today.

He had, he said, "pointed to the openings that could lead to future Swedish membership", and "in this situation, we could in the normal way investigate the advantages and disadvantages [of accession]."[43]

How can this apparent change in policy be understood? Was a conscious decision taken by the Social Democratic leadership, between Carlsson's two *Dagens Nyheter* articles, to alter its policy, in the light of the pressure it was under in the inter-party arena? Almost certainly not. Interviews with some of the actors involved suggest that confusion and misunderstanding were prevailing themes in this period. In the case of Carlsson's first article, the sub-editor's choice of headline may have sent quite a different signal from that the prime minister had intended. Such a view is supported by the contrasting inferences that were drawn from the article. Westerberg wondered publicly after the prime minister's first article whether it had actually opened the door to the possibility of Swedish membership; Bildt, meanwhile, drew precisely the opposite conclusion. Leading figures in the labour movement had similarly different impressions of the article. Some accounts suggest that it was actually written by Marita Ulvskog, Carlsson's press secretary and later a leading anti-membership campaigner, and was amended by Sten Andersson.[44] Other reports, gathered during interviews for this study, suggest that Carlsson himself was furious at the headline, and later admitted to colleagues that his own wording had perhaps been too cautious, making his statement unclear. Certainly, it is hard to see, after a close reading of the articles that Carlsson wrote in late May and early July, any great shift of position.

In fact, caution did indeed remain the Social Democratic watchword on European policy. "The obstacles to Swedish membership would be strengthened if foreign and security policy co-operation [in the EC] was deepened," Carlsson wrote in his July article, and he also claimed agree-

43 Ingvar Carlsson, "'EG hinder kan undanröjas'", *Dagens Nyheter* July 5th 1990. Emphasis in original.
44 *Dagens Nyheter*, June 13th 1991, cited in Gustavsson (1988), p.64.

ment with other party leaders "that there was not now cause to raise the membership question; that it was a discussion we could return to gradually; and that the time for this would be when we know more about how the EC is developing, and when we know more about what is happening in Europe as a whole".[45]

Doubtless this caution reflected SAP's internal problems. By late June LO was threatening more strikes. With the triennial party congress to be held in September and an election due the following year, Social Democratic leaders were desperate not to inflame internal dissent further, and a more open stance towards EC membership risked doing just that. As a former Moderate leader noted, with Carlsson's government facing serious opposition within SAP on tax reform, the Öresund bridge, nuclear power and allowing television advertising, the party leadership hoped to keep the European issue off the political agenda in the run-up to the congress—especially as "the EC opponents in the governing party are the opponents [of the leadership] in all the other questions as well."[46] An early pointer to how explosive the question had the potential to be within the party was an article by two LO economists, warning Carlsson against an application.[47] On May 27th—the day on which Carlsson published his first article in *Dagens Nyheter*—a Sifo opinion poll suggested that 51 per cent of voters were in favour of Swedish membership of the EC; but that figure included a significantly smaller proportion of Social Democratic supporters.[48]

Towards the policy change: September-October 1990

In the approach to SAP's congress in Gothenburg in September, pressure on the SAP leadership continued to grow. In the wider context of Swedish politics, a consensus was emerging about the desirability of joining the European Monetary System (EMS). The governor of the Riksbank publicly supported such a step. Meanwhile, on September 12th, the "two-plus-four"

---

45 Carlsson (1994), "'EG hinder kan undanröjas'".
46 Ulf Adelsohn, "Carlssons motiv: en lugn kongress", *Dagens Nyheter* June 10th 1990.
47 Dan Andersson and P.-O. Edin, "'Säg nej till EG, Carlsson!'", *Dagens Nyheter* June 24th 1997.
48 Of all the parties' supporters at that time. 82 per cent of Moderate voters were in favour, 71 per cent of Liberal voters, 33 per cent of Centre voters, 38 per cent of Social Democratic voters, 21 per cent of Communist Left voters and 40 per cent of Green voters. *Dagens Nyheter* May 28th 1990.

talks in Moscow concluded in agreement on German reunification the following month. Then, on September 15th, there was another significant external development. The European Commission published the Community's formal position in the EEA talks. It made clear that, contrary to the EFTA countries' hopes, there was no possibility of the Community's allowing non-member states any say in its decision-making procedures. This obviously made Sweden's most realistic alternative to EC membership that much less attractive.

Pressure was also growing from pro-EC trade unionists. Gudmund Larsson had been irritated by his colleagues' Eurosceptical article, fearing that their views might be seen as LO's. He decided that the best response would be to have the confederation take some position on the matter before the SAP congress in September. He thus instigated a new memorandum, which was subsequently upgraded into a report, to the LO European committee on Sweden's situation vis-à-vis the EC. Just before its publication, a week before the opening of the SAP congress, the Paper-Workers' Union held its own congress. The tone of this gathering was (to many observers) surprisingly pro-EC, and this was epitomised by a speech delivered by LO's chair, Malm. "We cannot run our economy today in the way we did before," he said. "In other words, we pay a political price for standing outside the EC...If, for example, we cannot come to an agreement between the EC and EFTA, the door must be kept open for a discussion about membership [of the Community]."[49]

The report presented by Larsson's committee—which was unanimously accepted by LO's executive—was in favour of a successful conclusion to the EEA negotiations, and took no hard and fast line on EC membership. But support for that eventuality was not ruled out, if the international situation, the course of the Community's development and Sweden's terms of accession were all favourable. This was an important first step for LO towards embracing the idea that membership might be a desirable objective. Moreover, in his capacity as a member of the Social Democratic Executive Committee, Malm was active in advocating that the party take a similar stance at its congress.

In the event, the Social Democratic congress was a fractious affair; but the European issue was not a cause of controversy. Indeed, there was scarcely a debate on relations with the Community. In his keynote address, Carlsson talked of five "visions" for Europe's future: an all-European se-

---

[49] *LO-tidningen* September 7th 1990.

curity order, including the Soviet Union and perhaps based on the CSCE; reduced "military thinking", to take the pressure off Swedish neutrality; "that the EC's foreign and defence policy work takes such a direction that the need for a common defence policy falls away"; "that Europe develops into a citizens' home—not a new arena for big business and primitive capitalism"; and that Europe look outward to address the problems of the wider world. The Danish Social Democratic leader, Svend Auken, suggested that Jacques Delors, the president of the European Commission, admired the example the Nordic welfare model could set for Europe. Auken told his Swedish colleagues that their country was needed at the European negotiating table. While agreeing that the EEA was the immediate priority, he also urged SAP: "Don't dally too long. The European train has been moving fast in recent years."[50]

But other issues were considered more important by congress. The economic crisis in Sweden had worsened. Feldt's successor as finance minister, Allan Larsson, presented the situation bluntly. The way inflation had got out of control was illustrated by his description of how nominal wage costs had risen by 28-30 per cent in the previous three years, while real wages (that is, taking account of price rises) had grown by just 2 per cent. "It is bloody serious now," he said. "We must stop inflation."[51] Other issues, especially the Öresund bridge, were all more controversial than Europe. The leadership had set aside three hours for such a discussion on the EC, but it was not needed; out of 700-800 motions submitted to the congress, only two referred to European policy.[52] The leadership was simply accorded the authority to deal with the matter as it saw fit. Congress adopted a position similar to the one LO had taken: the EEA was to be the priority, but "If the European states are brought together, and war in Europe becomes impossible to contemplate, or if co-operation in the EC does not need to occur in the area of defence policy—then Swedish membership of the EC might not be unthinkable."[53]

Contemporary figures in the party leadership insist that this did *not* mean that they wanted no discussion at congress. Toreson, the party secretary, emphasised to journalists before it that the leadership wanted a debate. Yet it is hard not to see the lack of debate on Europe as a success for

---

[50] *Aktuellt i Politiken* October 5th 1990.
[51] *Aktuellt i Politiken* October 5th 1990.
[52] One, from the local party in Alingsås, urged the party not to apply for membership.
[53] *Aktuellt i Politiken* October 5th 1990.

the party leadership, such were its difficulties with other divisive questions. The National Executive could be understood as continuing to prioritise its strategy in the intra-party arena rather than that in the inter-party arena. Moreover, even if the fingerprints of the finance minister, Allan Larsson, can be detected in the subtle drafting of the congress resolution that kept the leadership's options open on the EC, the trade minister, Anita Gradin had said to congress that to debate membership then would "not be very meaningful", and that the party's attention was needed "to solve the problems we face in the short term"—clearly implying that the issue was being kicked into the long grass.[54]

Away from the congress, however, pressure on the government to move Sweden in a more pro-EC direction was continuing to grow. By mid-September Andersson, the foreign minister, was musing about a decision on Swedish accession by 1993, security conditions permitting. At the start of October Carlsson presented his administration's plan for 1990-91, in which the prime minister suggested that "In a Europe with a new peace order, and in which the blocs have disappeared, it should be possible to reconcile [Swedish] EC membership with continued neutrality."[55] Again, though, this did nothing to alleviate criticism of its position. *Dagens Nyheter*, for example, derided what it saw as "a government without a policy".[56] Nor had the party's congress improved its public image: Sifo puts its support at just 32 per cent for September. But still there seemed little to suggest that the government would announce its intention to apply for membership within the next year, never mind before the month was out. So what prompted this culmination to a policy change that had already been fairly rapid?

The currency crisis

By the autumn the government admitted that inflation was expected to exceed 11 per cent by December, and that the 1989-90 budget deficit of SKr40 billion would probably expand by half again the following year. Short-term interest rates were already the highest in Western Europe, but it seemed inevitable that they would have to rise further. The OECD, among others, forecast a rapid rise in unemployment (which, in an overheated

---

[54] Gustavsson (1998), pp.176-77.
[55] *Dagens Nyheter* October 3rd 1990.
[56] *Dagens Nyheter*, "Regering utan politik", October 4th 1990.

economy, still stood at only 1.5 per cent). Industrial production had fallen by 3 per cent in the first half of 1990 alone. The government's plans for a wage freeze, or "zero agreement", were also steadfastly opposed by the trade unions, worsening relations within the labour movement still further. But it was then that the Social Democratic leadership was forced to switch the focus of its political strategy once again, as Sweden was hit by a classic currency crisis.

From around September 20th the Stockholm bourse began to fall steeply. A big financial company, Nyckeln, announced that it was suspending payment of dividends on its shares, and its competitors immediately followed suit.[57] This preceded an emergency package of measures, announced by the government on October 2nd. The opposition, the trade unions, the media and, most importantly, the financial markets were signally unimpressed with the package, and—crucially—heavy pressure began to build on the krona. Emphatic denials by Allan Larsson that the currency was about to be devalued were undermined not only by Sweden's history of devaluing, but also by leading economists, including, P.-O. Edin of LO and Assar Lindbeck, either predicting or openly advocating such a step. A Danish bank's newsletter, which forecast devaluation, triggered a huge outflow of capital from Sweden—perhaps SKr12 billion in a week.[58]

On October 12th the Riksbank raised short-term interest rates by 2 per cent, and by another 3 per cent six days later. The same day the government was forced to announce yet another crisis package, and the cabinet spent the following week composing it. It seems that on October 21st Carlsson and Allan Larsson agreed that it would contain a pledge to apply for Community membership, and they informed the cabinet that evening.[59] The EC application was duly included in the crisis package—which also featured cuts amounting to 1.5 per cent of public spending, in sick pay, public administration, education and rail transport—when it was published as an official letter to the Riksdag on October 26th. One commentator remarked how such a momentous decision had been announced—almost as a footnote in the crisis package.

---

57 Mikael Stigandal, "The Swedish Model: Renaissance or Retrenchment?", *Renewal* vol. 3, no. 1, 1995, p.19.
58 *Nordisk Kontakt* 15/16, 1990.
59 Gustavsson (1988), p.177.

Explaining the policy change

The interpretation offered here is that two environmental changes—the success of the campaign by the Moderates and Liberals to use the European issue to win electoral support, and the short-term effects of economic and financial crisis—forced the Social Democratic leadership to refocus its strategy, and to shift its priority from achieving its goals in the intra-party arena to achieving its goals in the inter-party arena. Certainly, this view is compatible with contemporaneous observations. The day after the package's presentation to the Riksdag, *Dagens Nyheter* declared that the decision "was a reaction to an acute flood of money and a component of an economic crisis programme, the history books will say". *Aftonbladet*, a pro-SAP tabloid, asked pertinently:

> What has happened in six months to provoke such a sudden change? The prime minister pointed mainly to the encouraging process of peace and democratisation in Europe...However, the positive statement on the EC will mostly be interpreted as a signal to industry and the finance markets.

Some in the labour movement argued subsequently that the policy change was coming anyway, and that the financial crisis had only a marginal effect on the timing of its announcement.[60] In his recent study of the change in European policy, Gustavsson suggests that Carlsson had gradually become convinced of the necessity of Swedish accession to the Community, a process that began with the fall of communism in Europe. He argues further that the counsel of his policy advisors, the lobbying of influential business representatives and, perhaps above all, contact with fellow social democratic leaders from elsewhere in Europe were collectively decisive in convincing Carlsson that the Community could be a platform for, rather than an obstruction to, his ideological preferences. This may be true. Equally, it may be an example of the type of retrospective rationalisation that Gustavsson himself identifies in other Swedish political actors.[61] Arguably, given the methodological problems involved in trying to ascertain the "real" motives of political actors, it may be of limited use to try.

---

[60] Toreson says that even Sten Andersson had been persuaded, after much analysis from the Foreign Ministry, that Swedish neutrality was not necessarily incompatible with membership of the Community. Interview with the author, June 1997.

[61] Gustavsson (1998), pp.114-15.

Where Gustavsson is surely right is in his emphasis on the key agency in the policy change, and the structural conditions that allowed it to happen. For whatever reason, Carlsson does seem to have become increasingly sympathetic to EC membership. So too had his old friend and political ally, Allan Larsson, and together the prime minister and finance minister—the latter traditionally the second-highest-ranking cabinet member—formed a very powerful pro-Community alliance in the party leadership. Certainly, there was still opposition to the change in SAP, not least because of the way it was implemented.[62] Yet, as Gustavsson shows, Carlsson and Larsson *exploited the currency crisis* to secure their preference, because they tied the application to the EC directly to it. First, they used the atmosphere of crisis to make the government and the party less cautious about contemplating radical options, and perhaps even to portray alternatives to the EC announcement as even more painful for the labour movement.[63] Second, by including the pledge in a package designed to ease pressure on the krona, they brought the issue of Community membership under the aegis of the finance minister, thus marginalising two of the less pro-EC cabinet members, Andersson and Gradin.[64]

It does not require too much counterfactual reasoning to infer, as Gustavsson does, that "With better economic prospects...the government would have retained its policy of non-EC membership."[65] The reaction of other EFTA countries offers some interesting evidence in this respect. In the previous few years, contact with other Nordic governments had been extremely close.[66] The Finnish government, whose security perspective was so intertwined with Sweden's, and the Danish government, the Nordic region's representative in the EC, had been consulted especially closely

---

62 When the crisis package was put to the Social Democratic parliamentary group on October 23rd, there was surprisingly little dissent about the EC section, but objections were raised by two MPs. One, Berndt Ekholm, was in favour of the application, but his fear was that more time should be made for consultation and preparing opinion in SAP.

63 Allan Larsson had apparently told Social Democratic MPs the week before that cuts of SKr25 billion were required, rather than the SKr15 billion eventually delivered. *Nordisk Kontakt* 15/16, 1990.

64 Gustavsson (1998), pp.182-86.

65 Gustavsson (1998), p.193.

66 Nordic co-operation is formalised in the Nordic Council, which was established in 1952; Finland joined three years later. It comprises delegates from the Swedish, Danish, Finnish, Norwegian and Icelandic parliaments, plus those of the autonomous territories of Greenland, the Faroes and Åland. The Nordic Council of Ministers provides a forum for government ministers to meet.

about developments. Regular contact had also been maintained with Norway. Moreover, the Nordic social democratic parties had also collaborated extensively, with the general secretary of each meeting four or five times a year. Yet the Finnish government was particularly taken aback when Sweden announced its intention to apply, and became more so when, a few days later, Andersson floated the idea of a joint Nordic application. The Finnish social-affairs minister, Tuulikki Hämäläinen, complained:

> We are surprised at Sten Andersson's remarks and very irritated about the fact that the Swedish government did not consult its neighbours before taking this drastic step. The time is not right. His statement could affect EFTA's negotiations with the EC negatively.[67]

Her president, Mauno Koivisto, reportedly accused the Swedes of "yet again" making decisions with implications for other Nordic states without consulting them. Finnish ministers suggested openly that the idea had more to do with bolstering the confidence of Swedish business than Nordic co-operation.[68] It seems most likely that the other Nordic countries were not informed of the Swedish announcement beforehand for the simple reason that the decision to make it had been taken so quickly.

Nothing, meanwhile, had happened to reduce the salience of the main obstacles to Swedish accession that Carlsson had stipulated in both his articles during the summer, and in his speech to the party congress. Indeed, it seemed increasingly likely that the EC would shortly adopt much more developed co-operation in the fields of foreign and security policy. Just two days after the crisis package was unveiled, the European Council agreed in Rome to convene intergovernmental conferences on deeper political and monetary union. As late as October 10th Gradin, the Swedish trade minister, had been telling the Riksdag that as far as the government's European policy was concerned, "In the short and medium term there is no alternative to the EFTA route." In the same debate, Andersson stated that "In a longer perspective, Swedish membership can become an issue." As *Dagens Nyheter* commented wryly on November 1st, "We now have a

---

[67] *Financial Times* November 1st 1990.
[68] *Nordisk Kontakt* 15-16/90, November 1990. There were also repercussions in Norway. The day after the Swedish government announced its intention to apply, the Eurosceptical Centre Party withdrew its support for the non-socialist government, whose Conservative prime minister was known to be sympathetic to the idea of a Norwegian EC application. The government was forced to resign.

definition of what this government means by a longer perspective"—namely, about a fortnight. It continued: "We stand with a cobbled-together European policy before a wondering world."

The fundamental question, then, is *why* the Social Democratic leadership—the party leader and his finance minister above all—used the opportunity presented by the crisis in this way. It may well be that EC membership accorded with their ideological preferences. But it also conforms with our analytical model of the leadership's interests and behaviour in this situation. It was in the leadership's interests, on one hand, to avoid risking the unity of the party, which a change in European policy would make likely. But on the other hand, it was also in the leadership's interests to maximise the party's support among voters. At the time, the EC was relatively popular. But a bigger incentive was probably the desire to avoid having the government's basic economic strategy collapse, which is what a devaluation of the krona would have represented, and which would undoubtedly have been electorally disastrous. This economic strategy depended on maintaining an anti-inflationary, "non-accommodating" monetary policy, and the means through which this, in turn, was to be implemented was through a fixed exchange-rate. It is discussed, along with Social Democratic approaches to monetary policy in general, in chapter 7. For now we need only be aware that the short-term survival of this strategy came to override all other objectives in autumn 1990, and that it became intertwined with the prospect of EC membership.

Conclusions

The aim of this chapter has been threefold. First, it has sought to demonstrate the plausibility of the notion that Sweden's economic difficulties were not so much the result of exclusion from the EC, but rather due to domestic problems. Whatever the precise combination of long-term flaws in the Swedish model of political economy, fundamental changes in production and markets, and short-term policy errors, it seems likely the roots of the economic crisis were either home-grown or global. In other words, non-membership of the Community was not the cause of the crisis, and membership was not necessarily the long-term solution. That Sweden's economy has scarcely been in vibrant health during the rest of the 1990s, even after accession to the EC, might be taken as further evidence of this position, albeit certainly not conclusive. The point is, though, that while there were strong additional arguments for EC membership, the Social Democratic government's decision to apply for it in 1990 was by no means inevitable.

The second aim of the chapter, and the reason why so much detail about the events that led up to the announcement of the application has been included, has been to convey the flavour of the debate that surrounded European policy during 1990. At first, Social Democratic leaders' main concern was the intra-party arena. They wanted to avoid aggravating disunity within the party, which would have debilitated their chances of attaining their goal—that is, of retaining office through maximising electoral support. Later, however, they were forced to adjust their priorities, and turn their attention to the inter-party arena, and thus put intra-party harmony at risk. This was due to two environmental changes: first, the success enjoyed by the Moderates and Liberals in exploiting the tide in favour of the EC among Swedish voters (a tide caused largely by external political developments); and, second, the onset of an acute economic and financial crisis. This latter change made it necessary to reassure the international financial markets about the government's economic policy. Achieving this objective became the government's immediate priority, as failure could have had a series of disastrous political and economic consequences. Thus, Social Democratic policy towards the EC changed concomitantly, moving from reluctance to address it, to a more open but cautious stance, and finally to announcing the application. In short, the change was only indirectly connected to developments in Sweden's economic and geopolitical environment, and was much more directly attributable to the short-term political interests of the Social Democratic leadership.

This in turn brings us to the chapter's third objective, which relates to our basic research question, about why the party has been so divided by the issue of Sweden's participation in European integration. It suggests that the economic and political crisis of 1989-90 contrived to pitch the short-term interests of the leadership against the factors that had underpinned a Eurosceptical disposition in SAP over a much longer period. The rapidity of the volte-face, the lack of overwhelmingly convincing reasons for it and, not least, its association with Social Democratic policy failure all created a well of resistance to it in the party. *The manner of the decision to change European policy, then, was crucial to the party's subsequent division over the issue.* As the following chapter will illustrate, the leadership's attempts to manage the divide may then have entrenched it.

6 Managing a divided party, 1990-94

In this chapter, we aim to establish a fuller picture of the nature of the Social Democratic division over EU membership. This will be done through analysis of the party leadership's management of the issue. The main part of the chapter will investigate the leadership's strategy for containing the effects of this potentially explosive question, looking at the strategy's formulation, its content and its implementation. A fuller depiction of the two issue-groups within SAP—their birth, structures, leaderships and finances—will be presented. From the time in autumn 1990 when the Social Democratic leadership announced its intention to submit a Swedish application for EC membership, it was aware of the decision's potentially divisive effects on the unity of the party and the labour movement. In many ways, the issue-management strategy that the leadership adopted succeeded in mitigating these effects. There may, however, have been a longer-term price to be paid.

Formulation of the strategy

Arguably, there were three areas in which SAP's strategy improved on, for example, that pursued by Labour in Norway before that country's referendum in 1972: in avoiding complacency; in setting the date of the referendum; and in its treatment of the anti-accession lobby within its party.

Of the three, complacency was the least of SAP's problems. The 1980 referendum on nuclear power offered a salutary lesson. As we have seen, Social Democrats had been divided by the question. The advisory nature of that referendum, and the fact that none of the three options on offer to the voter came close to securing a majority of votes, allowed the issue to be parried by the leadership, forestalling an open split in SAP. Sweden's EU referendum was also advisory, not mandatory, to its

133

government.[1] But after long and fraught negotiations with the Union over Sweden's terms of entry, it would have been impossible for any government to prevaricate on this issue: either Sweden would be a member, or it would not. In any case, and as we have also seen, previous referendums on European integration in Norway and Denmark were more than sufficient testimony to the political dangers the question could pose to a Nordic social democratic party. During the early 1970s DNA made two (as it turned out) disastrous miscalculations. First, it overestimated its ability to rally its members and supporters to the cause of EC membership.[2] Second, it had similarly overestimated its chances of keeping the issue from contaminating normal party politics. True, that was much harder in Norway because more parties (the Centre, Christian People's, Liberal, Socialist People's and Communist parties) were keen to make Europe a domestic issue for their own electoral purposes; in Sweden in the 1990s this was only true of the small Left and Green parties. Nevertheless, others' experience was more than sufficient to forewarn SAP of the perils the European issue held for a divided party.

As the events that ended the cold war unfolded in Europe and, in July 1989, Austria applied for EC membership, the first priority of SAP's leadership was to keep its options open. However, as the party formed the government at the time—it did not lose power to a non-socialist coalition until September 1991—party management could not be its sole, even its major, concern. Rather, its leaders were preoccupied with adapting Swedish official policy towards the EC. Amid rapidly changing external circumstances, this responsibility would have been a testing one for any Swedish government. Moreover (and as we saw in the previous chapter), Carlsson's administration was also under intense and increasing pressure from gathering domestic political and economic crises, which culminated in its pledge in October 1990 to apply to the Community. So as far as issue-management within the party was concerned, the leadership suddenly found itself facing almost a *fait accompli*: Sweden seemed headed inexorably for EC membership.

---

[1] There was no legal requirement for Sweden to hold a referendum on EU membership. The necessary adjustment to the constitution to facilitate accession needed only two parliamentary votes, one before and one after a national election. But it was soon clear that the matter was sufficiently controversial to make a referendum politically unavoidable if the decision, whether it was for Sweden to join or stay out, was to be perceived as legitimate.
[2] Allen (1979), p.102.

This was likely to arouse deep opposition from many Social Democrats. Yet, ironically, with hindsight these rapid developments may actually have been no bad thing from the point of view of SAP's leadership. It must already have seen that framing any strategy for the realisation of three relevant goals—implementation of its policy preferences (by then, Community membership), electoral success and party unity—would be a difficult business. Now, at least, one variable in the formulation of that strategy—the responsibility of determining the government's European policy—had quite suddenly been fixed. The party's internal issue-management strategy could now be devised in the knowledge that a Swedish application for Community membership had been all but submitted.[3]

From this point, then, the Social Democrats could begin the issue-management process. Just before Christmas 1990 the party's National Executive met to discuss its plan of action. On the initiative of the foreign minister, Sten Andersson, it was decided to appoint a working party from within the Social Democratic parliamentary group, whose job it would be to propose a strategy. At first, it was composed exclusively of members who were open to the idea of Sweden's joining the Community. Its chair was one of the keenest Social Democrats, Pierre Schori. But the chair of LO, Stig Malm, a member of the Executive Committee, soon pointed out the dangers of having such a one-sided committee being seen to decide the policy of the entire party. Andersson then became its chair. He was an elder statesman of the party, who had been its secretary between 1960 and 1982; the appointment of such a senior figure as chair of the working party symbolised the leadership's awareness of the EC issue's extreme sensitivity within the party. As we saw in the previous chapter, until late 1990 he had been doubtful about the desirability of joining the Community, and several other people of a sceptical disposition were also drafted onto the working party. Its 12 members included Anita Gradin, the trade minister; Gudmund Larsson and Bertil Jonsson, both of LO; Lena Klevenås, who had long been active in SAP's international campaigns on such issues as apartheid in South Africa and the Pinochet dictatorship in Chile; Mats Hellström; and Schori. Its secretary was Conny Fredriksson.

This balance of opinion within the working party was an essential part of its role in the leadership's issue-management. At this early stage, it contained no outright advocates of either joining or not joining the EC,

---

[3] The application was in fact submitted officially by Carlsson on July 1st 1991.

though most of its initial membership was sympathetic to the former, and several later campaigned actively on the Yes side. But the recruitment of more Eurosceptical members changed the character of the working party. It could no longer act as a *policy-making* committee, as it was unlikely that its members would be able to agree. On the other hand, it became increasingly valuable as a tool of *party management*. While there was always a majority in the working group that backed the leadership's pro-membership inclination, the Eurosceptical minority was large enough to prevent that element of the party feeling excluded from the decision-making process. Klevenås was later to be one of the most prominent Social Democratic No-sayers. Of the seven new members the working party acquired during 1992, two, Enn Kokk and Marita Ulvskog, were later to oppose Swedish accession, Ulvskog actively.

The timing of the referendum

With two political goals to juggle, the timing of the referendum in relation to the election was obviously of vital importance to SAP's management strategy. Setting a date involved balancing the requirements of several different priorities. After the economic and political disasters of 1990, the election of September 1991 was, in effect, discounted, and a bourgeois coalition of the Moderates, Liberals, Christian Democrats and the Centre Party duly took office after it. Many speculated that because the coalition's four constituent parties were dependent for a parliamentary majority on New Democracy, an unpredictable, populist newcomer to the Riksdag, it was unlikely to survive the full parliamentary term.[4] This proved not to be the case, however, and by 1993 it seemed likely that there would be no election in Sweden before the one scheduled for September 1994. With this date in mind, therefore, SAP's leadership could consider the date of the referendum.

For various reasons, many in the government parties wanted to hold the referendum before, or possibly simultaneously with, the election. The latter option was floated by the Moderates' secretary in October 1993, and although Bildt, the Moderate prime minister, quickly distanced himself from the suggestion, the motivation behind the government's support for

---

[4] For instance, Diane Sainsbury, "The 1991 Swedish Election: Protest, Fragmentation, and a Shift to the Right", *West European Politics* vol. 15, no. 2, p.166.

an early referendum was not hard to discern. The four applicant countries opened negotiations with the EC in February 1993, and March 1994 was set as the target date for their completion. Meanwhile, the economic situation that had contributed so much to the Social Democrats' defeat of 1991 was worsening further, and, as it presided over Sweden's worst recession since the 1930s, the coalition became increasingly unpopular with Swedish voters. The leaderships, at least, of all its four parties were in favour of EC membership. They believed that a successful conclusion of the negotiations, followed by a referendum that approved their result soon afterwards, might yet create a political momentum that they could use to retain power in the election of September 1994. Some Liberals, however, were also reviewing the option of forming a bridge between the blocs of Swedish politics, by defecting from the coalition and coming to some kind of electoral arrangement with SAP. A pretext for such a cross-bloc electoral alliance might have been the necessity of securing a Yes in a coming EU referendum. Thus, from the Moderates' viewpoint, a pre-election referendum would also have deprived the Liberals of any such pretext for defecting.

SAP's leaders, however, as they juggled the requirements of their three goals, were emphatically opposed to such an order of events. This was despite the possibility that in some ways it might have helped to achieve one of their objectives: with the EU issue decided, Social Democrats could have devoted all their energies to uniting in the 1994 election campaign. Indeed, this view was advanced by the anti-accession elements in SAP. They argued that the party might lose votes in the election to parties campaigning on an anti-EU platform (the Left and the Greens) if the question had not already been resolved. But from the party leadership's viewpoint, a pre-election referendum would have endangered at least two of its strategic goals, policy implementation and vote-maximisation.

First, Norwegian experiences suggested that it might actually do serious damage to the Social Democrats' electoral prospects. True, the most recent election in Norway had borne out the argument that votes might be lost to the anti-EC parties: in September 1993 the Norwegian Centre had received its best ever result, nearly tripling its share of the vote, by promising to vote against EU membership in the Norwegian parliament, even if a referendum approved it. But such difficulties were comfortably outweighed by the consequences of a pre-election referendum that, as we saw in chapter 2, Labour suffered in the 1973 election. The European debate within any divided party is bound to become most intense in the immediate run-up to the referendum; the incentive to display unity in the interests of winning a future election will inevitably grow weakest at that moment. Then, public exposure of the party's internal divisions will be most vivid—and the cost to the party's image is likely to be highest. This had certainly been the case for Labour in 1972-73. It thus made obvious sense for SAP to avoid an election for as long as it could after the referendum, so as to give the party as much time as possible to rebuild its internal harmony. In this light, holding the referendum right at the start of a new parliamentary term was ideal for the Social Democrats.

Moreover, a pre-election referendum would have seriously, perhaps fatally, jeopardised another of the leadership's objectives, a Yes to EU membership. It was widely assumed that Social Democratic supporters would be decisive in the referendum: even in SAP's disastrous electoral performance in 1991, they comprised 37.7 per cent of the vote. It was also acknowledged by leading Social Democrats that having to argue for the same outcome in the referendum as Bildt's party was their biggest problem in persuading SAP members and supporters to vote Yes—especially while Bildt remained prime minister. He was accused by Social Democrats of identifying his government's austerity measures with the need to prepare for "Europe". Carlsson, for example, declared that "The government blames its own ideological experiments on EC adaptation. So I am not at all surprised that opinion has changed [negatively] in the EC question."[5] With a Moderate-led government, SAP supporters would be less likely to accept that the Union could advance the cause of social democracy, and was not simply a free-market project—or, as one member of the Moderates' youth organisation put it, a place "where politics is given a few limited tasks in guaranteeing and underpinning the market economy".[6]

For that reason, the Social Democratic leadership was less than determined that the negotiations should be completed on time, as delay would rule out any chance of a pre-election referendum. Their slow

---

5 *Aktuellt i Politiken* September 10th 1993. Later Carlsson said openly that if his party's supporters were to be won over to membership, "we must show that issues like social-security, unemployment and the environment will be better solved if we are in the EU than if we stand outside. This plan is very badly wounded when the government goes out and says that we must cut taxes and rein back welfare if we join the EU. If the referendum had been held now, in June, as the Moderates wanted, I don't believe that we would have got a Yes." *Veckans Affärer* June 13th 1994.
6 *Dagens Nyheter* October 24th 1993.

progress drew complaints from Bildt in September 1993, but the previous month Carlsson had told *Göteborgs-Posten* that he doubted whether the timetable was realistic.[7] In September Hellström said that not only was a pre-election referendum "ruled out", but that because it was only really imperative that the applicants acceded by 1996 (when the Community's next intergovernmental conference was due to begin), rather than 1995 (the scheduled date of entry), the referendum could ideally be held as late as spring 1995.[8] Towards the end of 1993, Carlsson was emphasising that "To get as good a result as possible in the negotiations, each question must be dealt with very thoroughly, and that takes time."[9]

The negotiations began to pick up pace after agreement, just before Christmas 1993, on extending Sweden's right to maintain certain environmental standards, and to have *snus* (chewing tobacco) sold freely for some years after accession. Despite some late problems in the areas of agricultural, fisheries and transport policies, and on the precise size of the applicants' contributions to the EU budget, the talks between the EU and Austria, Finland and Sweden missed their deadline for conclusion by just a few hours, finishing in the early hours of March 2nd 1993. (Norway agreed its terms a fortnight later.) Some Moderates were still mooting the possibility of a referendum as early as June that year; Bildt had done so during a visit to Finland on January 15th. But in fact SAP's leadership was never in serious danger of having an early referendum foisted upon it. Even if Bildt had genuinely believed that a Yes to membership, to which he was very much committed, could be achieved in a pre-election referendum, it was clear that this would be impossible without the support of the Social Democratic leadership (if not the entire party) in the campaign. If Bildt had insisted on a referendum before the election, Carlsson might have been pushed into abandoning his multi-goal strategy and withdrawn his wholehearted support for a Yes vote. There was talk in Social Democratic circles of, *in extremis*, supporting neither a Yes nor a No, but some kind of middle line—much as the party leadership had done during the vote on nuclear power 14 years before.

Of course, Bildt also had to balance two goals his own, of victory in the election and a Yes in the referendum. While the strategy of SAP's leadership required the maximum separation of the two in the eyes of the

---

[7] Pedersen (1994), p.129.
[8] *Aktuellt i Politiken* September 17th 1993.
[9] *Svenska Dagbladet* November 28th 1993.

voters, his required their close association. Ultimately, however, it seems that he was sufficiently keen to keep the Social Democratic leadership on board the pro-accession side that he sacrificed an early referendum to ensure its support for a Yes. His desire to maintain a cross-party consensus with SAP on the issue—which occasionally caused disquiet within his coalition—was illustrated by his government's placid acceptance of a document containing 50 demands in the coming negotiations with the EC,[10] which SAP presented to it on the eve on the talks. The Moderate prime minister surely knew that without the wholehearted support of SAP's leadership, the prospect of Sweden joining the EU would be in real jeopardy. Three days after the conclusion of the membership negotiations, the leaders of Sweden's parliamentary parties met to discuss the date of the referendum. On March 18th it was announced that the vote would be held on November 13th 1994—nearly two months *after* the election.

Handling the anti-accession faction

The Social Democratic leadership's plan for dealing with the party's anti-EU element can be seen as a three-stage process. The first stage involved a strategic decision. The leadership had to decide how to approach a section that opposed its (likely) European policy (support for EC membership): to accommodate it, or to attempt to coerce it into toeing the leadership's line. Given that SAP's leaders opted for accommodation, the next two stages were essentially tactical. They concerned, respectively, the questions of how exactly to organise relations with Social Democratic Eurosceptics; and precisely when and how to come off the fence and subjugate the requirements of party unity to those of securing a Yes in the referendum.

---

10 SAP, *Socialdemokratin inför EG-förhandlingarna*, Politisk Redovisning Nummer 2 (Stockholm, SAP, 1992). These demands included that: Swedish nonalignment be retained; a decision on Swedish particpation in EMU be postponed until after 1996; the Swedish system of "openness" in drafting legislation be retained, and promoted in European law-making; a special scrutinising committee, like Denmark's, be established in the Riksdag; Swedish become an official Community language; the free-trade agreements between Sweden and the Baltic states be preserved; Sweden's participation in EC sanctions against third parties be preceded by a vote in the UN Security Council; Swedish regional policy be preserved; and collective agreements between employers and unions continue to apply to all firms, regardless of nationality, operating in Sweden.

Stage one: accommodation, not confrontation

By January 1991 the Social Democratic working party had agreed the leadership's initial strategy. As in Norway's Labour Party 20 years previously, the size of the anti-accession lobby within the party was considered simply too large for the leadership to demand that the minority accept the wishes of the majority, as on other policy issues (if, indeed, the leadership could command a majority in the party on the European issue, which was not certain). Any attempt to have done so would have been to risk provoking a serious rebellion against the leadership.

So, again like Labour before 1973, SAP as a whole was to take no official position on the merits or otherwise of EC membership. Instead, once more as Labour had done, the Social Democrats were to undertake a "neutral" study-campaign. The leader of one such group was later quoted as saying: "My job is not to influence the participants in any particular direction. Instead, we try to clarify some of the uncertainties many people have about the EU."[11] Local parties and affiliated clubs (local trade-union associations, for example) were involved, as were publishing groups associated with the Social Democrats, including Tiden and the Workers' Educational Association (ABF). An author and journalist, Göran Färm, whose previous work had included a biography of Carlsson, was commissioned to write a book setting out the arguments on both sides,[12] which appeared at the start of 1993. He was a supporter of EC membership, but his book was generally well received by members of the working party. Only two sceptics, including Klevenås, refused to accept its impartiality. This educational programme continued up to the referendum in September 1994. By 1993 the party estimated that 12,000 people had participated;[13] in mid-1994 the party secretary claimed that as many as 35,000 people had taken part.[14]

The policy of non-commitment was reconfirmed when, on January 19th 1993, the Social Democratic National Executive and parliamentary group met to discuss SAP's official position towards the EC. It was agreed, almost without dissent, to refrain from making any recommendation to the party's members and supporters before the conclusion of Sweden's

---

[11] *Aktuellt i Politiken* March 11th 1994.
[12] Göran Färm, *Sverige och EG. Det nya Europa—hot eller mojlighet?* (Stockholm, Utbildningsbrevskolan, 1993).
[13] SAP, *Verksamhetsberättelse 1990*-92 (Stockholm, SAP, 1993), p.221.
[14] *Aktuellt i Politiken* June 23rd 1994.

negotiations with the EC; and, most importantly, that there would be nothing obligatory for the party's members about the leadership's recommendation when it eventually came.

In many ways, this strategy conformed to the Swedish political tradition of scrupulous preparation of legislation and thorough investigation of its likely consequences. The Riksdag and its various committees debated the matter extensively, especially after the non-socialist coalition won power in September 1991. For instance, the Foreign Affairs Committee conducted a reassessment of Swedish security policy after the close of the cold war, and concluded that the country's policy of nonalignment was compatible with EU membership. The Constitutional Committee, meanwhile, asserted that the so-called "EEC paragraph", inserted into Sweden's constitution in the 1960s with an eye to possible entry, had to be expanded to facilitate membership, but that no explicit declaration of the supremacy of EC law over national law was needed. A commission decided that EU institutions were sufficiently democratic to be otherwise compatible with the constitution. The government also formed several commissions to investigate the consequences of accession. One, chaired by Olof Ruin, concerned Swedish democracy; another, under Sverker Åström, examined foreign policy. Both were published in early 1994.

Yet some criticised the Social Democrats' decision. Lindström, for example, describes it as "political abdication", which "released the party, indeed the entire labour movement, from closing ranks behind the leadership".[15] But for the Social Democratic leadership to have confronted the Eurosceptics would have been supremely dangerous. Elsewhere, Lindström concedes that "The invoking of party loyalty was tantamount to throwing a boomerang,"[16] and in de-linking the EU issue from the normal bounds of SAP's internal discipline, its leadership hoped to preclude the possibility of being struck by its own returning weapon. "Political abdication" might have somewhat weakened the Yes lobby across the spectrum of Swedish politics, but anything else would, in risking internal rebellion in SAP, have put the leadership's authority dangerously on the line—and thus risked Carlsson's other goal, winning the election. Besides, the policy of "wait and see" could be justified, not unreasonably, on the grounds that it would scarcely have helped Sweden's negotiators in

---

[15] Lindström (1994), p.72.
[16] Svåsand and Lindström (1996), p.209.

Brussels if the country's largest party had committed itself to supporting EC membership before the terms of accession had even been agreed.[17]

Party structures and relations with the Eurosceptics

Although the Norwegian Labour Party in the early 1970s and SAP in the early 1990s both chose the strategy of accommodation in their internal issue-management, aspects of its implementation differed. Indeed, in discussing the management of the division within the party, Carlsson stated openly that "We have learned from some of what happened in Norway in 1972."[18]

The leadership's strategy had a double-edged quality. On one hand, it shied from direct confrontation with anti-EU Social Democrats, so as to avoid putting its authority in the party at risk. True, as Lindström suggests, this might itself have undermined the leadership's authority; it did indeed amount to renouncing its responsibility for shaping party policy. But to some degree the problem was mitigated by the consistency of the leadership's position. The combination of cautious and qualified support for eventual membership, plus accommodation of those with different views on the EU, did not have the sort of insidious effect on morale within the party that other instances of more vacillating leadership have been seen to sow.[19] Social Democrats at least knew where they stood. Moreover, the risks of being seen to abdicate the responsibility of leadership were judged to be less than those inherent in provoking outright defiance by a section of the party. These conciliatory tactics led the leadership to debar the party's organisation from being deployed by either side in the campaign. Furthermore, unlike Trygve Bratteli, the Norwegian prime minister, in 1972, Carlsson conspicuously declined at any stage to stake his leadership on securing a Yes in the referendum, partly again to avoid having a significant element of Social Democratic voters being seen as challenging his authority by voting No.[20]

---

[17] See, for example, a leader in *Aktuellt i Politiken*, September 10th 1993.

[18] *Veckans Affärer* June 13th 1994.

[19] Allen (1979, p.93) postulates that the "Janus-faced" attempts to keep the support of both the pro- and anti-EC wings of his Centre Party by Per Borten, the Norwegian prime minister until March 1971, meant only that "Gradually but steadily, the last shreds of mutual confidence were worn away and the government's standing destroyed."

[20] Of course, de-linking the referendum from Carlsson's leadership must also be put in a broader political context. For every Eurosceptical Social Democratic voter that a threat of

And yet the dangerous but powerful weapon of party loyalty was not entirely abandoned. In de-linking the EU issue from its authority, the leadership may have chosen not to invoke loyalty directly; but the other side of that coin was the refusal of the leadership to be bound by a decision of the party congress on accession. The Centre Party, which was similarly divided, opted to make the result of its special congress binding on all members, the leadership included. In contrast, Carlsson stated that he personally would vote Yes whatever congress decided. Even though they were constantly reminded of their freedom to hold a different opinion, loyal Social Democrats were aware of precisely where their leader stood.

Meanwhile, the leadership's strategy also involved its taking incremental steps to secure its strength in the party, so that when the EU issue's dénouement did eventually arrive, the leadership's line could win maximum support. This process could be observed at the party's congress in Gothenburg in September 1993. There was hardly any opposition to continuing the wait-and-see policy, but the anti-EC lobby did object to the leadership's proposal to adjourn the congress and reconvene it, with the same delegates, after the government's negotiations with the EC had been concluded. The Eurosceptics wanted a completely new congress on the issue to be held, with delegates elected specially for it, and claimed that to adjourn the current congress for such a long period would violate the party's rules. "To judge by the statement from the party leadership, the conclusion of the negotiations will be significantly delayed and the timetable for the referendum put back," said the chair of Social Democrats Against the EC, Sten Johansson. "The option of adjourning congress should already have been ruled out for that reason."[21] The leadership argued that the existing delegates had been elected to formulate the whole range of Social Democratic policies, and individual policies should not be subjected to separate decisions. But adjournment also held a clear tactical advantage for the pro-accession side. Although anti-EC sentiment existed throughout the party, it was undoubtedly strongest at the grass-roots level. Delegates to a new congress on the EU question would have been chosen by their local organisations purely on the basis of their views on that issue, and the leadership believed that they would have included a greater proportion of anti-membership Social Democrats than delegates elected to

---

resignation might have brought back behind the leadership's line, a supporter of the non-socialist parties, especially the Centre and Christian Democrats, would probably have been persuaded to vote No so as to undermine a Social Democratic prime minister.

[21] *Dagens Nyheter* August 20th 1993.

a normal congress. In addition, local Social Democratic organisations and individual members would not be allowed to submit new motions to the reconvened congress, although they could make submissions to the official pro- and anti-accession party committees (see below).

In the event, the leadership got its way in Gothenburg: delegates voted 306 to 25 to adjourn their gathering (although a two-year time limit was placed on the adjournment). The margin of victory was impressive; but the complaints of the anti-accession side were understandable. In a separate decision, the congress voted to establish a standing committee to organise special congresses, to discuss both general policy and particular themes, between the main gatherings.[22] Johansson complained: "First congress decided to raise the possibility of extra congresses in order to increase the membership's influence within the party. Then it decided not to call an extra congress on the most important question."[23] But, in truth, the Gothenburg decision did not amount to a major confrontation between the leadership and the anti-accessionists, largely because the latter did not consider it worthwhile challenging the leadership openly on the issue. One leading anti-EU activist later conceded that a specially elected EU congress would still have supported the leadership's line, albeit by a smaller margin.[24]

The congress also reinforced the pro-EC side's representation in SAP's leading bodies. Two supporters of accession—one of them Mona Sahlin, since February 1991 the party secretary—were elected as full members of the Executive Committee. In addition, four more Yes-sayers, including Leif Blomberg (until recently the chair of the powerful Metal-Workers' Union) and Göran Persson were chosen as deputy members, without voting rights. Although at least one other non-voting member of the Executive Committee—Margareta Winberg, who, as chair of the Women's Federation, enjoyed *ex officio* membership—later became a leading anti-accession campaigner, the new recruits gave Carlsson a powerful pro-EC majority on it. It should be said that engineering this state of affairs was probably not a conscious priority for the leadership. The Social Democrats who had left the Executive Committee had also been favourable to EC membership, and Carlsson's chief priority was likely to have been simply to bring new blood, in the form of younger people and

---

[22] Reflecting the extension of the Swedish parliamentary term, the main Social Democratic congresses were thereafter to be held every four rather than three years.
[23] *Aktuellt i Politiken* September 24th 1993.
[24] Anders Ygeman, in an interview with the author, November 1994.

more women, into positions of power. But the new appointments at Gothenburg certainly had the effect of enhancing the Yes side's grip on the party hierarchy.

The trade unions were another vital constituency whose support the leadership hoped to secure. But the 23 trade-union federations comprising LO were also divided. The Euroscepticism of some members prevented LO's executive from taking a more enthusiastic position than the one put to and adopted by its congress in September 1991, which backed an application for membership but refrained from actually supporting membership itself. Meanwhile, Malm, who had initially been sceptical about closer relations with the EC, but who (as we saw in the previous chapter) had become more open to the idea in 1990, seemed to be becoming more Eurosceptical again. The property scandal that forced him to resign at the end of 1992 was a relief not only to SAP's leadership—his criticism of the Social Democratic government had been highly damaging to it in 1990-91—but also provided the opportunity for Bertil Jonsson, who was more enthusiastic about EC membership, to take over as LO chair. Again, his views on the issue are unlikely to have played much part in his elevation. Nevertheless, another important position in the labour movement was now filled by a Yes-sayer. Like SAP and LO, the Social Democratic Women's Federation, the Brotherhood Movement, Social Democratic Students and SSU all opted to delay their recommendations on accession until after Bildt's government had finished negotiating its terms. SSU, however, unlike the party as a whole, opted to send newly elected delegates to a special conference on the issue.

Perhaps as important as any aspect in the leadership's strategy was that the opponents of accession were accorded a certain legitimacy in the party, and thus kept where they were visible. This was in contrast to what had happened in Norway two decades before. There, the anti-accessionists had been given the scope to organise covertly; a secret campaign to obtain the signatures of party members in support of a more Eurosceptical Labour policy during 1971-72, ready to be invoked when the issue eventually came into the open, was a good example of this. In January 1972 anti-EC elements in the party had formed a "Labour Movement's Information Committee Against Norwegian Membership of the EEC". This was quite against the party's rules proscribing internal factions; yet such was its strength that the leadership could not move against it—which thereby undermined the leadership's authority and strengthened the confidence of the anti-EC lobby.[25] This was not an error SAP's leaders made. Although unwilling to confront anti-accession Social Democrats, the leadership was determined not to push them underground, and thus away from its influence. The anti-EC element was thus afforded an official status within the party. SAP's own rules against internal factions were, if not officially suspended, circumnavigated. Two committees were instituted, one to put the case for EC membership, the other to put the case against.

Anti-EC Social Democrats, as we have seen, faced numerous disadvantages compared to the supporters of Swedish membership: the latter included the party's leadership, dominated the upper echelons of its hierarchy and possessed the strength to influence decisively the timing of the referendum. However, with this weakness came the compensation of not having the leadership's responsibility for maintaining party unity. Thus, while the Yes side was having to tread carefully, especially before the conclusion of the government's negotiations with the Community, the No side could begin to organise. Social Democrats Against the EC had been formed at an SSU gathering a few months before SAP's Gothenburg congress. "Social Democratic Alternative to the EC", as it was initially known, was to provide another option to those who felt uncomfortable in the "No to the EC" movement, which was already in existence but which was considered by some in SAP to be too extreme generally, and too radically green in particular.

Social Democrats Against the EC was inaugurated in January 1993 and adopted its constitution the following March, by which time it claimed 300 paying members. By the Gothenburg congress it was a well-organised issue-group within the party, with representatives in all SAP's municipal branches. It had an articulate chair in Sten Johansson, who had been head of Statistics Sweden before resigning when Bildt's non-socialist government took power and taking up a chair in political science at Stockholm University. Its 26-member executive included such auspicious Social Democratic figures as Rudolf Meidner; Sören Wibe, a professor of economics at Umeå University; Inga Thorsson, a former ambassador; Einar von Bredow, who had finished in 1989 a 20-year career as foreign correspondent on a Swedish television current-affairs programme, *Aktuellt*; and two Social Democratic MPs, including Klevenås. Its secretary was

---

[25] As Allen asserts, "By the time Labour returned to office in March 1971 the internal opposition was already too large to be disciplined, let alone suppressed." Allen (1979), p.105.

Turanka Mülenbock. An early product of its organisation was a book, *Det nya riket?* (The New Country?), an anthology of 24 Social Democratic authors' arguments against Swedish accession to the EC. A newsletter, *Socialdemokratisk Europainformation*, was distributed.

Finance was a problem, though. Initially, the group had asked SAP for SKr100,000 and LO for SKr50,000, but had been turned down by both. Although ABF helped in the dissemination of its material as part of the party's educational programme, only later, just before the Gothenburg congress, did it receive a portion of the government's so-called "Fälldin money",[26] which was shared between the umbrella group on each side of the EC debate. In the anti-accession umbrella group, the "People's Movement Against the EU", the Social Democratic anti-accessionists joined a colourful range of other organisations opposing Swedish membership.[27] The People's Movement received SKr26.5m, of which SKr2.5m went to Social Democrats Against the EU. Nevertheless, the group could afford only two full-time employees; two others, based in Malmö and Gothenburg, were given month-long contracts preceding the referendum. All other workers were either volunteers or on government job-creation schemes.

When the two-committee policy was agreed at Gothenburg, Social Democrats Against the EC assumed the mantle of SAP's official anti-accession committee. In contrast, it was the adoption of this policy that prompted the formation of Social Democrats For the EC.[28] SAP's leaders did not repeat Labour's reluctance in 1972 to put heavyweight figures on its pro-accession committee: Pierre Schori, Mats Hellström, Mona Sahlin and Leif Blomberg were all recruited, as was the former finance minister, Allan Larsson. Social Democrats For the EC was chaired by Ines Uusmann, an MP and a member of the Women's Federation's executive. The group's committee was later joined by Sten Andersson, Bertil Jonsson and the chair of the Seamen's Union, Anders Lindström. Meanwhile, Odd Engström, described by *Veckans Affärer*, a Swedish news magazine, as

---

[26] This fund was so called because Thorbjörn Fälldin, a former Centre Party prime minister, was in charge of its administration.

[27] There were 26 groups represented in the People's Movement, including the Greens, the Left Party, members of all the other major parties, trade unionists, Friends of the Earth, "Free-Market No to the EU", Nordic unionists and "Youth with Allergies" (*sic*).

[28] It held its inaugural press conference on September 5th 1993, a fortnight before the Gothenburg congress, in the same Stockholm building, ABF House, in which Social Democrats Against the EC had their headquarters.

"the Social Democrats' trouble-shooter",[29] became active in "Yes to Europe", a non-party campaign group, in order to dilute its image as an organisation dominated by Liberals and Moderates.

While Social Democrats For the EC/EU aimed to win members from the grass-roots of the labour movement, other groups were also active in trying to persuade Social Democratic members and supporters of the merits of EC/EU membership. "Wage-Earners for Europe" had been launched some years previously, but was reactivated a few weeks before the Gothenburg congress. Anders Lindström was its chair and Gudmund Larsson its vice-chair; its executive comprised members of different trade unions. The closeness of its links with Social Democrats For the EU can be estimated by the fact that the two groups' headquarters shared not only the same building but also the same floor, and that during the referendum campaign they had the same secretary. In addition, some members of Social Democratic Students launched a campaign for a "Radical Yes!", and published a newsletter of the same name (*Radikalt Ja!*).

Social Democrats For the EU denied that it had large financial support: like its adversaries within SAP, it claimed at its launch to have received SKr250,000 of the Fälldin money and, in addition, SKr11,000 from private donations. Of the public money allocated for the referendum campaign, the pro-accession umbrella organisation, "Yes to the EU", received SKr22m (the People's Movement received slightly more because it contained more groups).[30] Its secretary claimed that Social Democratic group's share made possible the employment of about 30 people nation-wide, but in the light of the financial plight of Social Democrats Against the EU it seems unlikely that funding was not also received from other sources.[31] Yet it is arguable that the Yes side's chief advantage in the internal Social Democratic debate was not so much its material where-withal as the people it attracted to its cause, with their campaign experience and (in some cases) public profiles. At Gothenburg it was agreed that

---

29 *Veckans Affärer* January 26th 1994.

30 As well as Yes to Europe, Social Democrats For the EU and Wage-Earners For Europe, other organisations belonging to Yes to the EU included the green-tinged "Network For Europe" and the business-funded "Industry's EU Facts".

31 Indeed, Esaisson reports that, in the final fortnight of the campaign, Yes to the EU spent SKr5m on billboard advertising—the equivalent of all the state subventions it received during the entire campaign. Peter Esaisson, "Kampanj på sparlåga", in Mikael Gilljam and Sören Holmberg (eds), *Ett knappt ja till EU. Väljarna och folkomröstning 1994* (Stockholm, Norstedts juridik, 1996), p.33.

any party employee who wanted to campaign for either side could claim a month's paid leave before the referendum to do so. At least two people, and perhaps as many as 20,[32] from the party's national headquarters in Stockholm did so, and joined the Yes side. None from there joined the No side. This, one leading member of Social Democrats Against the EU privately argued, perhaps reflected party workers' clear understanding of the outcome their bosses in the leadership hoped to see.

Coming off the fence

In Sweden in 1993-94, as in Norway in 1971-72, the SAP and DNA leaderships had little option but to allow the anti-accession side a head start in campaigning. The Yes side could only campaign effectively once the leadership had committed itself wholeheartedly to membership, and it could only do that once the country's terms of accession were known. To do otherwise would have invited accusations not only of presumptuousness but also of undermining the bargaining position of Sweden's negotiators.

Of course, it would additionally have forced the leaderships to confront their anti-accession elements before it was really necessary. Labour had organised a long pre-referendum campaign in Norway in 1972-73: the referendum was held nearly nine months after the country's terms of accession had been settled and five months after the party had officially committed itself, at a special congress, to backing membership. Yet, partly because of complacency about its ability to turn opinion in its favour, especially among Labour supporters, the party's leadership had not used the time before the referendum effectively.[33] In a much shorter pre-referendum period—the campaign could only get underway properly once the September 1994 election was over—SAP's leaders could not afford similar passivity. They were helped by the favourable reception of Sweden's terms of accession to the EU, agreed in March 1994. Uusmann emphasised to Social Democrats that membership would not threaten collective agreements made between Swedish employers and trade unions, that security co-operation would be voluntary and that the Swedish state's monopoly in the retail of alcohol would be retained. The trade unions, too, were generally approving. Even Sten Johansson welcomed the outcome of

---

[32] Esaisson, "Kampanj på sparlåga" (1996), p.37.
[33] Allen (1979), p.166.

the talks, although he maintained that they were irrelevant to the fundamental issues that Swedish membership of the EU involved.[34]

The leadership decided to reconvene the Gothenburg congress to
decide SAP's recommendation to its members and supporters on June
18th-19th 1994, in Stockholm. Two days before, the National Executive
listened to presentations from Johansson and Uusmann, representing each
side of Social Democratic opinion. Despite the Euroscepticism that opinion
polls continued to show among Social Democratic supporters, the National
Executive voted 33 to none to support a Yes; there were not even any
abstentions, only a few absentees. The result of the congress itself was
scarcely in doubt, and despite some impassioned speeches the leadership's
victory was clear-cut: 232 delegates voted recommend a Yes, 103 favoured
a No. The vote came just a week after the Austrian electorate had voted
resoundingly to approve that country's terms of membership. Moreover,
elsewhere in the Nordic region, DNA and the Finnish Social Democratic
Party voted during the same weekend to support their countries' accession.
The previous week even Iceland's Social Democratic foreign minister had
mooted the possibility of his country's joining the Union if all the other
Nordic countries did so. Throughout the region the tide seemed to be
flowing the way of those advocating EU membership.

SAP's leadership had secured the endorsement it wanted. But,
crucially, it still had to accommodate the party's significant Eurosceptical
element if any meaningful party unity was to be preserved in the run-up to
the election. So it was magnanimous in victory: the leadership continued to
set great emphasis on the understanding that, despite the congress's
decision, Social Democrats could continue to oppose Swedish accession,
and contradict their leaders' advice, with a clear conscience. Even as the
editor of the party newspaper declared that its editorials would be
following the leadership's pro-accession line, he reassured his readers that,
"Naturally, a recommendation should not in any way be regarded as laying
down how a good Social Democrat should vote. A recommendation is a
recommendation, not a directive."[35] The message that it was equally
legitimate for a Social Democrat to be for or against EU membership was
constantly reiterated by the party leadership.

It was reinforced by the decisions of the groups within SAP and
the labour movement. The Women's Federation, the Brotherhood

---

[34] *Aktuellt i Politiken* March 11th 1994.
[35] Ove Andersson, "En rekommendation", *Aktuellt i Politiken* June 23rd 1994.

Movement and, most importantly, LO, all agreed to take no position. Bertil Jonsson, LO's chair, had promised to support whatever line its executive took, despite his association with Social Democrats For the EU. But Sweden's trade unions were divided. Those representing workers in export-orientated industries tended to support EU membership. For example, members of the Metal-Workers' and Paper-Workers' unions were especially active within Wage-Earners for Europe, and the Electricians' Union was also in favour. On the other hand, those that were not concerned with manufacturing or exports, and especially those in the public sector, tended to be much more sceptical. For instance, the only trade union to hold a ballot of its membership on the issue, the Transport Workers' Union, recorded a vote of three to one against EC membership. The congress of the Retail Workers' Union also voted against. And when, in June 1994, the congress of the Municipal Workers' Union decided to take no official position, and urged LO to do the same, it became likely that the confederation would have to equivocate in some way. Nevertheless, when LO's executive met to thrash out its position, a fortnight before SAP's congress reconvened in Stockholm, the gathering was a heated one.[36] Even a neutrally worded statement about promoting workers' interests in the EU debate was the subject of intense disagreement. The decision not to take a position was explained by Bertil Jonsson's statement afterwards: "It is extremely important", he said, "to keep LO together."[37]

Only the party's youth movements took unequivocal stances on the issue. A Social Democratic Students' congress voted 49 to 9 in favour of Swedish EU membership. SSU, however, voted 131 to 117 to oppose it, despite ten of its 13-member executive being in favour.

Outcome and conclusions

The issue-management strategy implemented by the Social Democratic leadership can be summarised as accommodation of opposition, combined with steadfastness about the leadership's objectives. In many ways, it can be considered a success.

---

[36] Its discussion was described by a participant as "rowdy...full of threats and protests against one side or the other". *LO-tidningen* June 17th 1994.
[37] *Aktuellt i Politiken* June 10th 1994.

The election of September 1994 resulted in a resounding victory for the party. Although SAP fell short of gaining an overall parliamentary majority, its 45.3 per cent of the vote was easily enough to oust Bildt's government and to form one of its own. Two months later the Swedish electorate voted to approve the country's accession to the EU—a narrow but decisive victory for the Yes side. Carlsson's twin objectives had thus been achieved. By the beginning of 1995 he was in office, and Sweden was a member state of the Union. For months before the referendum, opinion polls had indicated that a large plurality of Social Democratic voters had swung against Swedish accession. In November 1993 Statistics Sweden estimated that 52.3 per cent of SAP sympathisers intended to vote No, with 31.0 per cent uncertain and just 15.7 per cent planning to vote Yes. By May 1994 these figures had not changed much, standing at 47.4 per cent, 33.0 per cent and 18.7 per cent respectively.[38] Even an exit poll for the September 1994 election, the first survey since 1992 to predict a Yes in the referendum, still suggested that 52.4 per cent of Social Democratic voters intended to vote No, to 38.6 per cent intending to vote Yes.[39] From this position, half of Social Democratic voters ultimately took their leaders' advice to vote Yes.

The contributory factors in this double success can be identified. Issue-management was taken seriously. Senior party figures were detailed to devising and implementing the leadership's strategy. The lessons offered by the tribulations of SAP's sister parties in Norway and Denmark over the question of European integration, plus those elsewhere in Europe, were more than enough to dispel any complacency among Social Democratic leaders about the potential for internal damage to the party. Second, the timing of the referendum was, with hindsight, ideal for SAP.

Most importantly, the party leadership can be said—again, with hindsight—to have balanced its relations with Social Democratic Eurosceptics successfully. The combination of accommodating the anti-accessionists, while simultaneously securing crucial positions of power within the labour movement for the pro-membership side, seems to have been judged correctly. The two-committee system announced at the Gothenburg congress gave the No side official status within SAP. This kept it within the leadership's sphere of influence, and within the

---

38 Statistiska Centralbyrån (Statistics Sweden), press release, June 9th 1994. Cited in *Veckans Affärer*, June 13th 1994.
39 See the Valu 94 survey in *Dagens Nyheter* September 19th 1994.

constraints of all Social Democrats' obligation of loyalty towards their leaders. Had the anti-membership faction been driven underground, this obligation might well have been weakened, making more likely open rebellion within SAP, which in turn could have had severe electoral consequences for the party. In avoiding confrontation with the anti-accessionists for as long as possible, the leadership chose not to put its authority at risk. Yet when the campaigning for a Yes in the referendum could be put off no longer, SAP's leadership had secured for itself a position from which it could then campaign effectively. Having the Gothenburg congress reconvene to decide the EU issue, rather than holding a new congress, was a good example of this.

The leadership was also prepared to campaign fairly ruthlessly for a Yes if necessary—as, in the days leading up to the vote, polls suggested it was. Three days before the referendum the leader of the Municipal Workers' Union, Lillemor Arvidsson, announced that she had decided to vote Yes. This was a great boost for supporters of EU membership. Hers was Sweden's biggest trade union, with 660,000 members. Of these, 535,000 were women, and polls had consistently shown that women were more Eurosceptical than men. Pressure on her to state publicly that she would support accession had undoubtedly been applied by the party leadership. Furthermore, the tacit understanding that Social Democratic No-sayers thought they had with the leadership, that the latter would not resort to appealing to the loyalty of SAP voters, was broken in the campaign's closing stages. A week before polling day, Carlsson and Sahlin published a joint article in *Aftonbladet*, which concluded: "Those who trust us and want to give us the best means to take Sweden out of the crisis and halt the dismantling of welfare—should vote Yes to the EU."[40] Carlsson's finance minister, Göran Persson, also warned of higher interest rates and harsh cuts in public spending in the event of a No. After a generally good-natured internal debate, these last-minute tactics caused considerable bitterness among opponents of membership in the party and the labour movement. "Ingvar Carlsson played for high personal stakes when, not least in the campaign's final furlong, he put his own credibility on the line for EC membership," opined a leader in the LO newspaper. "It is possible that the 'trust-me' arguments helped some of the uncertain to decide in the days before polling day. But it left a rather bitter taste."[41]

---

40 Cited in Esaisson (1996). "Kampanj på sparlåga". p.41.
41 *LO-tidningen* November 18th 1994.

Of course, it is debatable how much choice the leadership had in formulating its strategy, and how much of it actually amounted to making a virtue out of necessity. Such an indulgent attitude towards the Social Democratic Eurosceptics was probably only possible because there was no faction within the party that sought to use the European issue as a means of challenging the leadership and promoting its own power. Accusations that Sten Johansson was presenting himself as an alternative party leader were never remotely credible. Indeed, he eschewed the possibility of organising the 5 per cent of the party's membership that, under its rules, could have demanded a party-wide ballot on the EU issue, lest it be seen as a general challenge to the leadership.[42] Nor could active campaigning for the leadership's policy have been delayed until two months before the referendum had most of the rest of the Swedish political establishment not been in favour of EU membership. Third, in many ways it was a blessing for the leadership to have been in opposition until just a few weeks before the referendum. It is hard to see how, given Sweden's economic situation, any government could have remained popular at that time; and an unpopular Social Democratic government would very probably have found it much harder to rally its usual supporters to vote Yes to the EU.

There is also the basic question of luck. Our interpretation of the strategy pursued by SAP's leadership would doubtless be very different if the referendum had gone the other way, producing a narrow No to EU membership—as it nearly did. In that case, Carlsson would have been blamed by others in the pro-accession camp for not campaigning sufficiently hard within his party for Social Democrats to vote Yes. In fact, in the last, tense week of the campaign, Bildt and the Liberal leader, Bengt Westerberg, levelled those very accusations at SAP's leader, as some polls suggested that—despite the clear Yes in the Finnish EU referendum less than a month previously—the No vote was edging ahead. But after the event, with a positive result secured, Carlsson's strategy had paid off. With nearly an entire four-year parliamentary term to run before the next election, SAP could hardly have had more time in which to repair the internal stresses sustained over the EU issue before having to face another election to the Riksdag.

Yet the strategy also had longer-term costs. Importantly, it was maintained after the election and referendum of 1994. To the surprise of some in the party, Carlsson appointed two anti-accessionists to his new

---

[42] Stated in an interview with the author, November 1993.

cabinet. Margareta Winberg, who had stridently denounced the CAP,[43] was made minister of agriculture, and Marita Ulvskog became minister for the civil service. The Eurosceptics' continuing accommodation was also illustrated by the complex manner in which the party chose to fight the European elections in September 1995, in which voters could choose between pro- and anti-EU Social Democratic lists.[44] The party's score was disastrous. Moreover, by suggesting that, even after the referendum, Social Democratic Euroscepticism was just as valid as support for the leadership's line, the strategy served to *legitimise* and thus strengthen the opposition to that line. By mid-1997 a senior pro-EU figure in the labour movement, who acknowledged that the accommodation of the Eurosceptics before the referendum had "saved the party", argued that the policy had become damaging.[45] Indeed, the European elections debacle also *institutionalised* Euroscepticism in SAP in a more tangible, and ironic, way: Eurosceptical MEPs have access to large subventions from the Union, which helps to fund their organisation.[46] Although Social Democrats Against the EU was dissolved, its activists continued to liaise in a network of "Social Democratic EU-Critics". Sören Wibe, who became an MEP, was the group's chair.

In the longer term, then, the price of Carlsson's twin successes in 1994 could be significant for SAP. Accommodating the anti-EU element to such an extent, and thus institutionalising and legitimising it, might not have mattered much had the referendum of November 1994 really brought a conclusion to the debate. But, as other Scandinavian social democratic parties have discovered, European integration has a habit of continually producing new and divisive issues for national parties to handle. As we shall see in the following chapter, this is what happened after Sweden joined the Union in 1995. SAP may yet come to regret allowing this internal conflict to gain a foothold in the party.

---

[43] Margareta Winberg, "Dyrt, byråkratiskt, miljömassigt sämre—om EGs jordbrukspolitik!", in Lotta Gröning (ed.), *Det nya riket? 24 kritiska röster om Europa-Unionen* (Stockholm, Tidens förlag, 1993).

[44] Two Social Democratic lists were offered to voters in each of five regions. The top 12 names on all the lists were the same, but their order was different: in each region, one list gave higher placing to candidates who were more positive towards the EU, while the other gave priority to those of a more sceptical disposition.

[45] Personal interview with the author, June 1997.

[46] Together with the Left Party, Social Democratic EU-Critics also issue an English-language newsletter, *Progressive EU Opponents*. This is also based in the European Parliament, and is distributed to sympathisers throughout the Union.

7 EMU rears its head

The rocky road to the single currency

During the campaign that preceded the Swedish EU referendum in 1994, the question of economic and monetary union between the member states was not given a terribly high profile. This was despite the country's terms of membership, which implied inexorability about participation in the single currency. The Treaty on European Union, signed by the then 12 member states at Maastricht in 1991, envisaged a three-stage, strictly timetabled progression to EMU, in which countries would complete the liberalisation of capital movements within the Union, harden their currencies' exchange-rate parities in the EMS and then finally—so long as they passed certain macroeconomic qualifications, or "convergence criteria"—hand over the right to manage monetary policy to a European System of Central Banks, with the ECB at its apex. If a majority of member states satisfied the criteria, EMU would begin in 1997. If not, the single currency would come about two years later, on January 1st 1999, even if only a minority passed the criteria. Two states, Britain and Denmark, negotiated "opt-out" clauses, which permitted them to stand back from stage 3—the irrevocable locking of exchange-rates and the transfer of monetary decision-making. Sweden, however, like the other new member states, had no such provision. The project's automaticity, which its chief proponents considered so crucial if it was to be realised, applied—in most legal opinions, at least—as much to the new members as to all but two of the others.

There were two main reasons for the relative lack of debate about EMU before the Swedish referendum. First, it suited the major parties. Given the difficulties they were having in persuading the electorate to approve the level of supranational integration required by basic EU membership, they naturally preferred to avoid drawing attention to something that would imply a much higher level. Bildt, for example, referred to EMU as a "highly technical issue". Social Democratic elites, meanwhile, when they did discuss it, insisted that no state could be forced to participate against its will, and that Sweden would decide whether it should join the single currency in a separate decision at the appropriate time. Some alluded to the final right of approval that the German and Dutch parliaments had unilaterally arrogated to themselves; others suggested that the convergence criteria could be changed, making them more responsive to "real" economic conditions (such as unemployment), once Sweden was inside the EU. There were those at the time, not least in Social Democrats Against the EU, who considered these last hopes to be unrealistic—a view that since then has been borne out. But there was another reason why Swedish politicians could relegate such an important issue as EMU to the sidelines in the referendum campaign. By late 1994, despite the fixed timetable, the launch of the single currency looked highly uncertain.

The recession of the early 1990s had contributed to low inflation and interest rates throughout the EU, thereby helping to satisfy two convergence criteria (see figure 7.1). But it had also damaged the public finances of the member states to the extent that only Luxembourg satisfied the criteria relating to public debt and budget deficits. Indeed, throughout 1995 German economists and some in the SPD were mooting a delay in the project to allow states to repair their accounts, and in October that year the EU's council of economic and finance ministers, Ecofin, confirmed that 1999, not 1997, had become their target for starting EMU. Most dramatically, though, it seemed that a fifth criteria, requiring a member state's currency to have participated fully in the EMS and its exchange-rate mechanism (ERM) for two years prior to EMU's launch, would be even more problematic. Europe's currency markets had suffered extreme turbulence in 1992, which had forced Britain and Italy to suspend their currencies' ERM parties and allow them to float (downwards). In July and August 1993 renewed speculative attacks all but forced the entire system to be abandoned. The bands in which most currencies were allowed to fluctuate were widened significantly.

German reunification had led to a boom in that country, financed in large part by a burgeoning budget deficit. Ever watchful for inflationary consequences, the Bundesbank had tightened monetary policy. Other EU economies were far less buoyant. But because the D-mark was in effect the ERM's anchor, other members were forced to import German interest-rate levels in order to maintain exchange-rate parities. The financial markets were betting that such excessively tight monetary policies would become politically unsustainable, and in some cases they were proved right.

There was, then, a straightforward explanation for these crises: an economic shock that had thrown the German economic cycle out of kilter

Figure 7.1 The convergence criteria for EMU

Budget deficit	No more than 3% of GDP
Public debt	No more than 60% of GDP, unless "exceptional and temporary" or if "declining substantially and continuously", and comes "close to the reference value"
Price stability	Average inflation during the previous year of no more than 1.5 percentage points above the three best-performing member states in terms of price stability
Interest rates	Average long-term interest rates during the previous year of no more than 2 percentage points above the three best-performing member states in terms of price stability
Currency stability	No devaluation, violation of "normal" fluctuation bands or severe tension within exchange-rate mechanism

Source: Treaty on European Union (1991), articles 104c and 109j, plus relevant protocols.

with those of other EU member states. But in the light of previous attempts at exchange-rate co-operation in Western Europe during the 1970s, which had also been hampered by financial-market disruption, a sense that the goal of a single currency would always be beset by such problems hung over the EU. This despondency added to pessimism about further integration that had flowed from the Danish electorate's No to Maastricht in June 1992, plus the way in which the French referendum the following September had come desperately close to producing a similar response. Thus, Swedish politicians were tempted to assume that EMU could be dealt with at a later date—if, indeed, it became necessary at all. After the bruising their party had taken over Europe in 1990-94, Social Democratic leaders lost little sleep over that situation.

But EMU did not go away. As German interest-rate cuts allowed monetary policy to be eased throughout the EU, currency stability returned and economies recovered; meeting the convergence criteria began to look increasingly realistic to a growing number of states. It also became clear that, for the purposes of the exchange-rate criterion, the ERM had survived. A decision on Sweden's position began to look more and more inescapable for the Social Democratic government. Unfortunately for SAP, the decision on EMU would arguably have been hard for the party even if the baggage of its previous volte-face on EC membership could have been discounted.

The Swedish context

The single currency is, in the judgment of most observers, particularly in the six states that founded the EC, primarily a political project, with the paramount aim of binding united Germany into European political structures.[1] By definition, then, it involves considerable transfer of political authority to EU institutions, which would be difficult for some member states, notably Britain and the Scandinavian countries, to countenance. But it is, of course, also a question of economic pros and cons. The economic advantages of a single currency are clear: reduced exchange-rate uncertainty, which should encourage cross-border investment; preclusion of competitive devaluations, which might undermine member states' commitment to the single market; fewer conversion costs between currencies; exposure of relative prices to more transparent comparison throughout the union; and, perhaps above all, a monetary policy regime that, thanks to the ECB's design, is likely to deliver low inflation. On the other hand, the great sacrifice in EMU is a country's ability to administer monetary and exchange-rate policies tailored to its particular economic conditions. In a currency union, such policies take account of conditions in the union as a whole, and they may thus be inappropriate for one part at a given time.

The cost–benefit analysis of EMU is, like EU membership itself, different for each member state (and, arguably, parts of individual member states). For those whose economies are particularly integrated with each other, there is less to be gained from retaining monetary autonomy. To have their currencies depreciate would be to see relative prices of imports increase, very probably with damaging inflationary consequences. On the other hand, unilateral attempts to boost domestic demand without currency depreciation would suck in imports and cause balance-of-payments problems. This was the outcome of Mitterrand's attempt to reflate the French economy in 1981-83. Pressure on the franc became so intense that it had to

---

[1] See, for example, Francesco Giavazzi, "EMU—the key to the crucial German question", *Independent* May 2nd 1996; Lars Calmfors, Harry Flam, Nils Gottfries, Rutger Lindahl, Janne Haaland Matlary, Ewa Rabinowicz, Anders Vredin and Christina Nordh Berntsson, SOU 1996:158, pp.231-52.

be devalued in the EMS on three occasions by an aggregate 30 per cent.[2] Alternatives to exchange-rate pegging were, therefore, unpalatable for some EU states. But the downside of fixed exchange-rates was that, in effect, it left these states' monetary policies almost entirely beholden to that of the Bundesbank, as the D-mark—partly because of its impressive non-inflationary credentials, more because of the size of the German economy—anchored the ERM. So an ECB in which policy-making was *shared* with Germany might well have amounted to an *increase* in its neighbours' monetary autonomy compared to the status quo.[3]

Pegged exchange rates were by no means new for Sweden. During the 20th century the value of the krona had been maintained in relation to other currencies, except in two brief periods between the wars. After the Bretton Woods system collapsed in the early 1970s, Sweden had also participated in the Community's first attempt to co-ordinate the currencies of its members, the "snake in the tunnel", in March 1973. However, the experience was not a happy or a stable one: after two devaluations, the krona was withdrawn in August 1977 and returned instead to its previous peg, a basket of currencies that included the dollar.[4]

Moreover, even outside the EC, Sweden had found itself embroiled in the currency turbulence of 1992-93. In May 1991 the Social Democratic government had unilaterally pegged the krona to the European Currency Unit (ECU), a basket of all its member currencies. In September 1992, at the same time as the British pound and Italian lira fell out of the ERM, the Riksbank had raised overnight interest rates to 75 per cent and then to a staggering 500 per cent to maintain the krona's peg to the ECU. As a non-member state, Sweden had no legal recourse to request help from other member states in defending the value of its currency. But, agreeing that the Swedish economy could not stand another huge hike in interest rates, Bildt's government nevertheless pleaded with the EC's monetary commit-

---

[2] Kenneth Dyson, *Elusive Union: The Process of Economic and Monetary Union in Europe* (London, Longman, 1994), p.115.

[3] Luxembourg's position was an extreme but illustrative example. In the field of monetary policy, Luxembourg ccould only gain from EMU. With its own monetary policy decided by the Beglian central bank since a currency union in 1921, Luxembourg had no autonomy to sacrifice. See Nicholas Aylott, "The European Union: Widening, Deepening and the Interests of a Small Member-State", EPRU Working Paper No. 3/95 (Department of Government, University of Manchester, 1995).

[4] Daniel Gros and Niels Thygesen, *European Monetary Integration: From the European Monetary System to Monetary Union* (London, Longman, 1992), p.17.

tee for a public statement of support for the krona. The committee demurred. That was also the reply it received from the Social Democratic leadership when, after two previous cross-party crisis packages designed to steady the currency, the coalition suggested a third. By the time the government finally admitted defeat and let the krona float in November (which led to a rapid depreciation of around 20 per cent), perhaps SKr160 billion had flooded out of the country in a single week, and at least SKr16 billion in public money had been lost in trying to defend the currency.[5] The attempt to establish a new guideline for Swedish monetary policy, in the form of pegging the exchange-rate to the ECU, had failed.

Yet, in themselves, the consequences of this policy failure were not obviously bad. Defending the ECU peg had inflicted a savage shock on the economy: in 1990-93 GNP declined by about 5 per cent, industrial production fell by nearly 18 per cent, the banking system came close to collapse and real unemployment rose to around 14 per cent. When the peg was abandoned, however, the subsequent depreciation of the krona boosted the competitiveness of Swedish exporters, while the depths of the recession had squeezed inflation out of the economy. The new monetary policy guideline became not an exchange-rate target but, as in Britain, a simple inflation target, which the Riksbank was mandated to meet through monitoring a range of indicators and adjusting interest rates accordingly. The krona, meanwhile, continued to float. The upshot of the attempt to peg the Swedish currency in 1991-2—what might be called associate- or semi-membership of the ERM—had been traumatic, particularly regarding unemployment. In view of this, and also of the stability that the post-1992 monetary regime appeared to bring, it was not surprising that the Social Democratic government remained deeply reluctant to contemplate full membership of the ERM. In short, the advantages of monetary autonomy seemed to retain their attractiveness, even after Sweden's accession to the EU in 1995.

In October 1995 Persson, the Social Democratic finance minister, commissioned a panel of economists to investigate the consequences of EMU, with Lars Calmfors in its chair. Its recommendation, delivered in November 1996, was that Sweden should not join in the first wave, in 1999. This was for two main reasons. First, Sweden was more vulnerable to economic influences that might affect it differently to other member

---

5 Ewa Stenberg and Gunnar Örn, "De spelade bort sexton miljarder", *Dagens Nyheter* November 3rd 1996; see also Hinnford and Pierre (1998).

Table 7.1 Estimated relative contribution of different shocks to fluctuations in GNP (percentage of "total shock")

	Symmetric component %	Asymmetric component %
France	76.3	23.7
Belgium	74.8	25.2
Germany	73.4	26.6
Austria	70.2	29.8
Netherlands	69.8	30.2
Luxembourg	55.4	44.6
Denmark	26.5	73.5
Sweden	**18.9**	**81.1**
United Kingdom	12.3	87.7
Finland	6.5	93.5
Ireland	6.9	93.1

Note: Symmetric shocks are those that affect all relevant states in a roughly equal way; asymmetric shocks affect countries in different ways.
Source: Adapted from SOU 1996:158, p.108.

states, particularly those in the "D-mark zone"—that is, those with their economies most integrated with Germany's (see table 7.1). Moreover, Sweden was less capable of adapting to such asymmetric shocks than some other member states. The rigidity of most West European labour markets, the report argued, retarded their ability to respond to changes in economic conditions: for example, labour mobility was low, and nominal wages were hard to reduce. In Sweden's case, though, adaptiveness was especially lacking; the report was "quite pessimistic regarding the possibilities of bringing about a greater flexibility in nominal wage formation, with the aim of reducing the risk of disturbances in employment attendant to membership of the monetary union".[6] This was the second reason for non-participation in 1999. With wage costs currently being restrained so ineffectively in the Swedish labour market, it was better, the Calmfors report suggested, for the country to try to introduce a more flexible system while preserving the means to loosen national monetary policy if economic disturbances cut demand for labour.[7]

---

[6] SOU 1996:158, p.399.
[7] SOU 1996:158, pp.19-21.

Social Democratic calculation

The Calmfors report was a relief to the Social Democratic leadership. A recommendation that Sweden join EMU in 1999 would have placed huge pressure on the government; the suggestion that, on balance, it was better to wait diminished the immediacy of the issue. After all, a decision to keep out of the monetary union could always be changed later. Once made, however, it would be all but irreversible. Yet there were real costs for SAP in Sweden's standing aside, both economic and political. The Calmfors report acknowledged that the political costs of exclusion might be significant.[8] It also looked positively at future Swedish participation, in more propitious circumstances. The Moderates and Liberals were nevertheless arguing for entry in 1999; more importantly, so were some influential figures in SAP. In addition, external developments could not be disregarded. We saw in chapter 5 how the end of the cold war created a "Europhoria" in Swedish public opinion. There was little that could be described as euphoric about public attitudes to the euro, as Ecofin decided in autumn 1995 to christen the new currency,. Yet as it became increasingly apparent that EMU in 1999 would be "broad", embracing the majority of the member states, including Finland, some Swedes wondered whether there would be a price to pay for isolation from such a large currency bloc.

On June 3rd 1997 the Social Democratic National Executive announced that its recommendation to the party congress regarding monetary union was "No for now". This was, in effect, a declaration that the Swedish government had decided not to participate in EMU in 1999. But just as the party leadership's decision to apply for EC membership had not been inevitable in 1990, neither was standing back from EMU a foregone conclusion. EMU had powerful supporters, as well as opponents, in the labour movement. The same had applied in 1990 when the issue was EC membership. Why, then, did the decision in 1997 veer in a Eurosceptical direction, when the move of 1990 had taken the opposite path? Once again, an analysis is required of the leadership's three primary goals: *votes*, in competition with other parties; *unity*, internally and with the rest of the labour movement; and *policy* implementation, with the attendant requirements of alli-

---

[8] This was especially because Sweden would probably also be reluctant to participate in a second "inner core" of the Union, pertaining to foreign and security policy. SOU 1996:158, p.423.

ances in order to form parliamentary majorities. Each goal shall be examined in turn in relation to the decision on EMU.

EMU and the pursuit of votes

For reasons that we have seen, the stipulation that prospective EMU members had to keep their currencies in the ERM for two years preceding the launch of phase 3 was one constraint on Swedish Social Democratic leaders. They considered an important Ecofin meeting in April 1996 a victory, as it agreed that membership of a post-EMU version of the ERM, or "ERM II", would not be compulsory for member states that did not join the single currency at its launch. After the same meeting, Sweden's Social Democratic finance minister, Erik Åsbrink, insisted that, despite what the Maastricht treaty clearly stated, the criterion stipulating two years' ERM membership had been rendered obsolete by the turbulence in the currency markets of August 1993, after which the bands in which ERM currencies were allowed to fluctuate against each other had been extended to 15%. Åsbrink argued that this was in practice equivalent to floating, and that consequently the Maastricht criterion could be satisfied by a currency simply achieving a reasonable level of stability, rather than being formally part of the ERM.[9] Although he and his British counterpart were and remained in a minority of two on that question, Åsbrink's stance provided the fig leaf that SAP's leaders required to keep their options open with EMU.

---

[9] *Dagens Nyheter* April 14th 1996.

Table 7.2 Swedish public opinion and EMU, May 1997

Sweden should...	participate from 1999 %	participate later %	never participate %	don't know, uncertain %
Inclined to vote for...				
Moderates	45	20	20	15
Liberals	27	37	30	6
Greens	22	23	49	6
Social Democrats	**20**	**24**	**39**	**17**
Centre Party	20	21	47	11
Left Party	10	31	49	10
Christian Democrats	9	14	57	20
Self-employed	39	17	24	20
Salaried employee	36	24	25	15
Wage-earner	**22**	**23**	**42**	**13**
Private sector	33	27	29	11
Public sector	**23**	**19**	**40**	**18**
SACO	36	13	34	16
TCO	33	20	29	18
LO	**19**	**25**	**44**	**13**
Southern Sweden	27	25	34	14
Central Sweden	27	23	33	18
Northern Sweden	17	21	46	16
Big cities	31	19	32	18
Towns pop. > 3,000	25	33	37	15
Towns pop. < 3,000	22	29	33	16
Men	34	23	31	12
Women	18	25	38	19
15-29	25	20	36	19
30-49	30	22	35	13
50-64	26	29	32	14
65-	20	25	35	19

Note: Constituencies especially relevant to SAP are in bold.
Source: Sifo.

Indeed, it was probably the fiscal criteria that more powerfully constricted Swedish political elites' ability to nurture public support for EMU, particularly on the left of the spectrum. Around 1993, when the Swedish budget deficit ballooned to over 12% of GDP and the public debt to nearly 100%, it seemed very unlikely that Sweden would be able to pass the relevant tests of EMU membership. However, a combination of spending restraints and tax rises, amounting to around SKr126 billion, was implemented by the Social Democratic government after 1994, and it restored the state's financial position sufficiently for it to be more than adequate for EMU qualification. These measures, including in 1995 a politically sensitive reduction of unemployment benefit to 75% from 80% of previous earnings, were painful. They did nothing to diminish the association by Social Democratic supporters of economic decline and austerity, not to mention the abandonment of the traditional priority of full employment, with European integration.

It is true that public belt-tightening was equally harsh, if not harsher, elsewhere in the EU without affecting public support for the single currency. To assume a close connection between austerity and Euroscepticism also implies that Swedish voters were impervious to the quite reasonable insistence of their politicians, Social Democratic and non-socialist alike, that fiscal retrenchment was necessary regardless of Maastricht and the convergence criteria. Nevertheless, the context was different in Sweden. For all the historical, cultural and political reasons that we have examined in previous chapters, the prospect of giving up the krona was unpopular in its own right. The association of austerity with preparation for EMU made public-sector retrenchment and tax rises acceptable in some countries, most strikingly illustrated by the Italian government's introduction of a "tax for Europe". But in Sweden, such association arguably made such measures even more politically difficult. Contemporary psephological evidence suggests that suspicion of the euro was especially strong in parts of the electorate in which SAP was traditionally strong: the working class, LO members, the public sector, northern Sweden (see table 7.2). In announcing the decision of the party's National Executive, Åsbrink argued:

Table 7.3 Social Democratic voters and EMU, May 1997

"What do you think the Social Democratic leadership should recommend to the party congress regarding EMU?"	%
Join EMU in 1999	12
Decide a date after 1999 when Sweden should join	3
Sweden should join EMU at some time after 1999, but not decide a date	13
Wait before deciding whether to join EMU	35
Not go in at all	25
Uncertain, don't know	12

Source: Sifo.

> The decisive factor in the government's view that accession to the monetary union will not occur in the immediate term is political. I can state that support for accession on January 1st 1999 is lacking. This is clearly shown by numerous opinion polls and other indicators…Politicians must take citizens' worries seriously. It is not in the interest either of Sweden or of Europe to try to drive through Swedish participation in monetary union from the start when there is insufficient public support for it.[10]

Social Democratic supporters were clearly in favour of Sweden's staying out of EMU in 1999 (see table 7.3). Moreover, Eurosceptics in the labour movement could claim allies on the Calmfors commission and, significantly, in the media. *Svenska Dagbladet*, a conservative broadsheet, sided with the Calmfors verdict on EMU and the Swedish labour market. Even more important was the stance taken by the Social Democratic tabloid, *Aftonbladet*, which in May 1997 came out against monetary union on democratic grounds. It would have been a brave party leader, therefore, who had tried to persuade Social Democratic members and supporters of EMU's merits. Yet, arguably, that is what has happened in other EU countries, notably Austria, Germany and—perhaps most relevantly for Sweden—Finland, where determined governments have, according to one's point of view, either ignored or tried to lead electorates that have been even more suspicious of the single currency than Sweden's. Public opinion was an important constraint on any inclination the leadership may have had

---

[10] Erik Åsbrink, "Sverige och EMU—Anförande av finansminister Erik Åsbrink den 3 juni 1997 kl 18.30", Swedish Ministry of Finance, June 3rd 1997 (www.regeringen.se).

to join the single currency in 1999. But it is not a complete explanation of its declining to do so.

EMU and party unity

Even when the issue of EU membership had been decided in 1994, EMU simmered as a potentially serious threat to the party leadership, largely because it both aroused the suspicion of the electorate and because it rallied Social Democratic opponents of the general policies that the leadership was pursuing in the 1990s. Two years after Sweden acceded to the Union, it was no great surprise to hear frustration in the party at the austere economic strategy given voice by the person who had led Social Democrats Against the EU in the 1994 campaign.[11] Yet as we saw in chapter 4, it is an over-simplification to characterise the differences of opinion within SAP on EMU as left against right. The mosaic of views within the party is more complex than that. But just as those fault lines do exist, so they are related in a rough way to the divide over EMU.

The labour movement, monetary policy and EMU

In some ways, the euro did crystallise a long-standing division within the Swedish labour movement over monetary and exchange-rate policies. Theoretical discussion within SAP and LO usually assumed a fixed exchange-rate, but it was by no means an inviolable norm. Management of a counter-cyclical macroeconomic policy—the "first component of the Swedish model"[12]—occasionally required an adjustment in the krona's value. The devaluation of 1931, for example, was widely credited with saving Sweden from the worst effects of the Depression and, through the advantages of an undervalued currency, with contributing to the growth the country enjoyed in the following decade (even if the devaluation had in fact been undeliberate).[13] But this was the exception to the rule. Both the Rehn–Meidner plan and the EFO model of wage formation, discussed in

---

[11] Sten Johansson, "'Ekonomisk galenskap, Persson!'", *Dagens Nyheter* January 19th 1997.

[12] Lundberg (1985), p.5.

[13] Assar Lindbeck, *Swedish Economic Policy* (London, Macmillan, 1975), p.23; Lundberg (1985), p.9.

chapter 3, referred explicitly to a fixed value for the krona as part of mac-roeconomic "stabilisation" policy.

With the exchange-rate fixed, monetary and fiscal policies became the tools of demand-management, and in doing so they were informed by the need to maintain the balance of payments. If high domestic demand threatened to suck in too many imports, the exchange rate could not be used to adjust relative prices, and so the onus was on monetary and fiscal policies to restrain demand. Wage-formation was thus also disciplined, as pay rises that endangered competitiveness could not be accommodated and cushioned by a depreciating currency.

But several developments undermined this model. When, follow-ing the oil shocks of the 1970s, the exchange rate began to be adjusted more frequently, the peg's credibility was progressively undermined. Thus, the balance-of-payments constraints that the fixed exchange-rate had im-posed were loosened. At around the same time the public sector began to grow very quickly, wage inflation spiralled and monetary policy was insuf-ficiently tight to restrain either of these developments. A second change came in the 1980s, as financial markets were deregulated. Governments throughout Western Europe discovered the difficulty of combining capital mobility, exchange-rate stability and independent national monetary poli-cies.[14] But with Swedish monetary policy already loose, thanks to the re-fusal of its governments—Social Democratic and bourgeois alike—to use higher unemployment as a means of suppressing domestic demand, a vi-cious circle, a "devaluation cycle", began to form, with inflation raising costs, undermining competitiveness and putting further downward pressure on the exchange rate. In short, as the Lindbeck report acknowledged, "The framework for stabilization policy was transformed in a radical way."[15]

The economic crises in the 1980s and 1990s prompted at least three responses in the Swedish labour movement. First, there was the strat-egy of the "third way", designed and implemented by Feldt, like-minded Social Democrats and officials in the Finance Ministry. It was radical in that it borrowed from so many intellectual sources. Its centrepiece, the de-

---

14 Cf. Tommaso Padoa-Schioppa *et al*, *Efficiency, Stability and Equity: A Strategy for the Evolution of the Economic System of the European Community* (Oxford, Oxford University Press, 1987); Benjamin J. Cohen, "The Triad and the Unholy Trinity: Problems of Interna-tional Monetary Co-operation" (1993), reprinted in Jeffry A. Frieden and David A. Lake (eds), *International Political Economy: Perspectives on Global Power and Wealth*, 3rd ed. (London, Routledge, 1995).
15 Lindbeck *et al* (1994), p.230.

valuation of 1982, was neo-Keynesian; its emphasis on supply-side measures, especially in the labour market, combined with continued prioritisation of full employment, harked back to Rehn–Meidner; its insistence that inflation must be defeated, that the public sector must be streamlined and, above all, that private-sector profit and investment were the keys to "saving and working out of the crisis" was—not least in the view of some cabinet colleagues—infused with neo-liberal thinking.[16] But as we saw in chapter 5, the culmination of the "overheating crisis" in 1990 and the failure of the stop package marked the demise of Feldt's faction in SAP. The debate concerning EMU, therefore, can be seen as a contest between two successor intellectual groups.

Social Democrats inclined to sympathise with the thrust of Feldt's reforms, particularly on the supply side, became in a broad sense the "modernisers" within the party. They agreed that the tax burden could not continue to rise indefinitely, that there was scope for making the welfare state more responsive to consumer demands, and that the private sector was the vital engine of growth. Naturally, they were as disappointed as any in the party at the precipitous rise in unemployment in the early 1990s and the government's failure, after 1994, to make much progress in reducing it. They tended to agree, furthermore, that the labour market could be made more flexible. Where they parted company with the third way, though, particularly in its original form, was in their perceiving the impossibility of cutting unemployment through traditional demand-management. Incomes policies, which Feldt had relied upon to maintain the competitive advantage derived from devaluation, had failed.[17] With financial markets deregulated, Swedish firms had to be allowed greater profit margins if they were not to relocate abroad. The proportion of the national debt in foreign hands had reached the point at which further depreciation of the krona would make servicing that debt unsustainably expensive. In short, the modernisers could see no alternative to a monetary policy that made controlling inflation, not maintaining full employment, the absolute priority.

As we saw in chapter 5, it was this determination to maintain an anti-inflationary, non-accommodating monetary policy that had underpinned the desire of the Social Democratic leadership to announce the application for EC membership back in 1990. Carlsson in particular seems to

---

[16] Cf. Löwdin (1998), p.244-5, 283-91.

[17] Nils Elvander, "Incomes Policies in the Nordic Countries", *International Labour Review* vol. 129, no. 1, 1990, p.20; Ryner (1994), p.253.

have felt that the potentially disastrous electoral consequences of a collapse of his government's basic economic strategy, which a forced devaluation of the krona would have signalled, could only be warded off by
pledging accession to the Community. At that time, before the adoption of
the Maastricht treaty, there were no EC rules constraining a member state's
fiscal position, which a government might have used as a self-imposed
straitjacket. There was also no prospect of the krona's joining the EMS
before Sweden joined the EC, which would take years to negotiate. The
system's collective support for the parities of the currencies within it, was
open only to full EC members (as the Bildt government discovered in
1992). However, announcing the application for EC membership did send
a signal. Nearly all EC member states were committed to exchange-rate
co-operation in the EMS, and, through its public intention to join the EC,
and thus also the EMS, the Social Democratic leadership sought to declare
that competitive devaluations to maintain Swedish competitiveness were
over. This, it hoped, would send signals to the parties in its own labour
market: if wage costs rose too fast, they would henceforth not be bailed out
by devaluation (which made all Swedes poorer in real terms), but would
instead feel the consequences directly, through reduced profits (for business) and higher unemployment (for labour).

For some in SAP, EMU represented the means finally to secure
such a non-accommodating monetary policy. Credibility would be its main
advantage over other means of fighting inflation. Pegging the exchange-
rate had failed consistently since the 1970s, but full monetary union would
definitively end the devaluation cycle, and thus both ensure economic stability and reintroduce an external discipline into wage-formation. Whereas
the Calmfors report had argued for reform of wage-bargaining institutions
before Sweden joined the single currency, pro-EMU Social Democrats'
arguments flowed in the opposite direction: they hoped that early participation would actually create the conditions in which such reform might
succeed. Unemployment, meanwhile, would have to be tackled in other
ways, perhaps at the European level. As early as September 1995, a month
after announcing his intention to retire as party leader, Ingvar Carlsson
voiced such a position. He called for EMU's realisation in tandem with an
EU "employment union", which would co-ordinate mainly fiscal counter-
cyclical measures, as the only means of tackling European unemployment.
"It is not an option to maintain Swedish competitiveness with a continually
cheapened currency," he argued. "To keep open the path of devaluation is

to keep open the way for the collective wage-cuts and reduced purchasing power that a falling krona involves."[18]

In contrast, the "traditionalist" wing—that is, the Rehn–Meidner plan's architects and supporters—retains a vocal presence in the party. Sweden's main economic problem, its adherents insisted, was the old one of constraining demand during an upswing, and it had been caused by a failure to resolve another old dilemma: how to constrain corporate profits. The wage-earner funds had been LO's proposal for dealing with this problem, being intended, at least in Meidner's original design, to allow extra funds to be created for corporate investment, without the problems of class inequity and wage drift that higher profits would normally entail.[19] (Indeed, in explicitly seeking to increase investment through higher profits, Feldt's third way was directly at odds with one of the basic precepts of the Rehn–Meidner plan.[20]) In addition, the increase in organised labour's control of company shares might also have limited the tendency for Swedish capital to decamp overseas. (Precisely because it facilitated such capital flight, the financial-market deregulation that Feldt oversaw in the 1980s was another object of criticism from traditionalists.) "Today there is nothing left of the restrictive economic policy that gave stability," Meidner lamented in 1997. "There is nothing left of the idea of co-operation between state, trade unions and capital, and nothing of wage solidarity. Now there is nothing but naked profit and internationalisation."[21]

For anti-EMU traditionalists, the project represented much of what was wrong about "modernised" Social Democracy. Deregulated financial markets precluded democratic control of capital. The ECB's constitution, which prescribed its inflation-fighting remit, ruled out its use of monetary policy to maintain employment. What Europe needed, traditionalists argued, was a macroeconomic boost to demand. But because such a strategy had insufficient support elsewhere in Europe, and because the ECB's inde-

---

[18] Ingvar Carlsson, "'EMU för jobbens skull'", *Dagens Nyheter* September 10th 1995.

[19] According to their original design, presented by Meidner in 1976, every year all enterprises of a certain size would contribute shares, equivalent to 20% of the firm's profits, to a wage-earner fund. When the fund had acquired 20% of the firm's equity capital, it would be passed up to one of several LO-controlled national fund boards. Half the dividends would go towards investment, half towards trade-union purposes. Meidner suggested that trade-union control of most major companies could be achieved with 25-50 years. Esping-Andersen (1985), p.298.

[20] Pontusson (1992), pp.186-219.

[21] *Dagens Nyheter* May 4th 1997.

pendence from democratic control made it impervious to political influence in any case, signing up to monetary union would rule out a more expansionist policy in Sweden. Thus, continued autonomy in monetary policy, with provision for adjustable exchange-rates, was necessary.

Handling the EMU question in SAP

As with the decision to apply for EC membership, the political and economic
background to a crucial Social Democratic decision on European policy was
extremely salient. The mid-1990s were another difficult period for SAP.
Carlsson announced his retirement in August 1995, and his deputy, Mona
Sahlin, looked certain to become Sweden's first woman prime minister.
But she was exposed as having used a government credit card for personal
items, and then being tardy in repaying the money; she resigned from the
government the next month. As it happened, Sahlin was strongly pro-EU;
and, before the credit-card scandal broke, the person who publicly flirted
with the idea of challenging her was one of the ministers who had campaigned against Swedish accession in 1994, Margareta Winberg. On the
other hand, Winberg's supporters seemed more motivated by her generally
traditionalist credentials, rather than her Euroscepticism in particular.

Criticism of the man who was eventually persuaded to accept the
jobs of party chair and prime minister, Göran Persson, also focused largely
on questions of domestic economic policy, particularly the government's
priority of repairing the public finances, rather than on EMU or Europe. As
economic recovery eased the fiscal position, he attempted to mollify this
criticism through promising more resources for the public sector. Even
before he formally took over, he promised to restore unemployment benefit to 80 per cent of previous earnings and to halve the number of jobless
by 2000, an objective to which he pledged SKr50 billion in June 1996. In
April 1997 the government announced a four-year programme against unemployment, with SKr66 billion—what became known as the "Persson
money"—devoted to preserving public-sector services, creating 70,000
extra higher-education places and providing early-retirement incentives. A
further SKr8 billion was promised the following September. Even this did
not forestall a sustained campaign from LO to promote its interests, however. Mayday that year gave vent to an atmosphere between the two wings
of the labour movement that was more poisonous than anything since
1990. A series of demonstrations against government policy occurred in

Stockholm in the summer and autumn of 1996. Then, in September, in response to proposed labour-market reforms, which included a new time limit on receipt of unemployment benefit, LO threatened to withhold its SKr20bn annual donation to the Social Democrats—perhaps a fifth of the party's national income. The government humiliatingly backed down.

Yet for all the power in the party that LO displayed, it probably exercised only limited collective influence over the party's approach to EMU—not least because LO was itself divided. In April 1997 its chair, Bertil Jonsson, came down against Swedish membership in 1999, echoing the verdict of the Calmfors report, that such were the problems in Sweden's labour market that the option of devaluation to retain competitiveness had to be retained in the medium term. But, as they had about EU membership, some of LO's most powerful trade unions took very different views of the single currency. The leader of the Retail Workers' Union chaired Social Democratic EU-Critics; in April 1997 the Metal-Workers' Union backed Swedish participation in EMU "at an appropriate moment".

Given all this disagreement, it may be that the party leadership shied from tackling the question of EMU not because the issue itself was highly contested *per se*, but because it did not want to aggravate splits in an already divided party. But it is equally arguable that the single currency really does divide SAP in a way that is separable from other questions of domestic policy. Evidence for this thesis can be found in the events of autumn 1996 to spring 1997.

By mid-1996 the party leadership was sounding sceptical about EMU. In July Persson raised doubts, not about its economic aspects, but about its democratic basis; and in August the finance minister, Åsbrink, suggested that to say "no now", with the option of "yes later", would be a reasonable position for Sweden.[22] Then, just after Christmas, Persson declared that with a European central bank pursuing a "very strong" monetary policy and the stability pact restricting national governments' room for fiscal manoeuvre,[23] EMU could in fact lead to European-level tax and finance policies—which, he felt, would amount to the advent of a federation, "something completely different to the EU that the Swedish people,

---

[22] Erik Åsbrink, "'EMU-anslutning kan senareläggas'", *Dagens Nyheter* August 28th 1996.
[23] This pact was agreed, at Germany's behest, in autumn 1996. It sought to extend relevant convergence criteria (on budget deficits, for example) to member states even after the launch of EMU.

after a long and hard debate, said Yes to."[24] This seemed to shatter a cabinet truce on the issue. In early 1997, a few days after the Metal-Workers' leader had called for the government to take a lead in promoting EMU, the agriculture minister, Annika Åhnberg, became the first member of the cabinet to call openly for Swedish participation.[25] The same week the minister of commerce, Anders Sundström, argued equally forcefully that Sweden should wait and see how the monetary union functioned after 1999.[26] Junior ministers started to declare their opposition to or support for EMU in more strident terms. This open disagreement within the government was aggravated in February by the governing council of the Riksbank, half of whom had Social Democratic connections, unanimously repudiating the Calmfors commission and advocating Swedish accession to EMU from its launch.

These demonstrations of government disunity abruptly ended. Quite conceivably, Persson, seeing how they were draining his authority, insisted that discipline be restored. However, he may have been assisted by simultaneous and ostensibly unrelated political developments. These point to the third of our contemporary constraints on the Social Democratic leadership's handling of EMU.

EMU and policy implementation

By 1998 the Social Democrats had governed Sweden for 60 of the previous 77 years. But on only two occasions, in 1940 and 1968, had they won a parliamentary majority, and the first of these was in the context of a wartime national coalition. Other than that, since the first world war the Social Democrats have only twice, in 1936-39 and 1951-57, governed with another party, the Centre. In 1960-1973 and 1982-88 SAP could do without sharing office, as the fairly reliable support of the Left (and its previous Communist incarnations) was enough to muster parliamentary majorities. Thereafter, however, the two parties of the socialist bloc did not have a majority of seats in the Riksdag, and Social Democratic governments had to bargain with non-socialist parties from a weaker position than previously. At the same time as SAP had to look rightwards, disappointing

---

24 *Svenska Dagbladet* December 29th 1997.
25 *Svenska Dagbladet* January 21st 1997.
26 *Veckans Affärer* January 27th 1997.

some of its supporters, the Left discovered the potential of exploiting this disappointment, and became a less reliable ally.

Clearly, the Social Democratic leadership's actions regarding the single currency were constrained by the effects of Sweden's multi-party system. Horse-trading had become unavoidable, and Eurosceptical parties could insist on coolness towards the EU as a price for their support. This was certainly the case in SAP's relations with the Left and the Greens; both parties won their highest-ever national scores in the 1995 European elections on the back of strongly anti-EMU platforms. But it also applied vis-à-vis the Centre. Although, as we saw in chapter 5, its leadership decided to support joining the EU, its members and supporters were, like SAP's, divided by the issue. After 1995 the Centre became a third Eurosceptical party in the Riksdag. Also in 1995 its leadership began formal parliamentary co-operation, falling short of coalition, with the new Social Democratic government.

What were the terms of this pact? Its immediate manifestation was in a budget that raised taxes and cut spending to the tune of SKr15 billion. But its foremost component became clear nearly two years later, when the Social Democratic and Centre leaderships, together with the Left, agreed to phase out two nuclear reactors, thus finally beginning to put the verdict of the 1980 referendum into effect. The deal was struck against vociferous opposition from the Greens, who wanted faster decommissioning, and the Liberals, the Moderates and also many in SAP and LO, who feared that it might cause still more unemployment. The Centre's leader, Olof Johansson, considered the agreement a great victory; but it was not the only one he could claim in justifying his co-operation with the Social Democrats to his (often doubtful) party. As early as February 1996 Johansson had declared that the Centre's support for the minority Social Democratic government was conditional not just on its getting satisfaction over nuclear power, but also on a commitment that Sweden would not enter EMU without a referendum.[27] In June 1996 the Centre's convocation voted nearly two-to-one against the single currency.

Of course, SAP's leaders—understandably loth to promise a referendum on the euro after all their difficulties preceding the one on EU membership—need not have been held completely hostage on the issue by the Centre. If agreed with the other bourgeois parties, a more pro-EMU stance would have commanded a parliamentary majority. But three consid-

---

27 *Dagens Nyheter* February 6th 1996.

erations held the Social Democrats back from that course. First, such a cross-bloc deal might not have been sufficient, because transferring more power to the EU without changing the constitution required a qualified parliamentary majority; and in May the smallest non-socialist parliamentary party, the Christian Democrats, opted to oppose joining EMU in 1999. This highlighted the second obstacle facing the Social Democratic leadership, that of maintaining party unity. The Christian Democrats' decision was enough to spark an outbreak of dissent within SAP: four leading figures declared that there were now enough Eurosceptical Social Democratic MPs to complete a blocking minority if the government tried to join monetary union without holding a referendum or amending the constitution.[28] With opinion in the party so febrile, any deal on EMU with either the Liberals or (especially) the Moderates would have caused enormous strains within the labour movement. Given its troubled recent relations with LO, the Social Democratic leadership hardly wanted to risk worsening them further.

Perhaps most importantly, however, there was also a third consideration. To have a realistic chance of forming an alternative government, the non-socialist parties needed the support of the Centre, as (despite several strains) they had maintained during Bildt's coalition of 1991-94. But nuclear power and EMU represented the main differences between, on one hand, the Centre and, on the other, the Liberals and the Moderates. If SAP conceded to the Centre's demands on those issues, not only would the two parties' pact be sealed, it would also lock the Centre into a deal that effectively burnt its bridges with the Liberals and the Moderates—greatly reducing the possibility of a non-socialist alternative government.

In this light, the end to the public airing of differences between pro- and anti-EMU Social Democratic ministers in early 1997 becomes more explicable. First, at the beginning of February, there was the deal on nuclear power. Then, in April, Johansson, speaking at a meeting organised by a Eurosceptical group, issued a clear political threat. "If the Social Democrats say Yes to EMU," he declared, "the consequence would be that the possibilities for [the Centre's] co-operating with them would be made

---

28 Kenth Pettersson, Bengt-Ola Ryttar. Lena Sandlin and Sören Wibe, "'Vi stoppar EMU-inträdet'", *Dagens Nyheter* May 19th 1996. Opinion was divided as to whether the part of the constitution dealing with the right to issue currency would require amendment in the event of Sweden's joining EMU.

harder."[29] "I am convinced that there was a misunderstanding," said Persson, in reference to Johansson's gambit[30]—doubtless perfectly aware that the tougher the Centre leader's anti-EMU rhetoric, the greater the distance would become between him and the other non-socialist parties. As the debate on monetary union intensified, in both SAP and Swedish politics generally, the Centre's vice-chair, in a detailed policy statement, made explicit his party's demand for a referendum before any attempt to accede to EMU.[31] It came as no surprise, then, when the Social Democratic leadership finally declared its hand and announced a policy of "No for now" to EMU at the beginning of June.

Conclusions

The Social Democratic leadership's decision on EMU was ratified by the party congress the following September, after only a short, though impassioned debate. There was no vote. In December the Riksdag confirmed the Swedish position.

The political conditions prevailing when the decision to stand back from EMU were, as we have seen, different to those applying when, during the economic crisis of 1990, the Social Democratic leadership had taken the decision to seek EC membership. In 1990 public opinion had been relatively well disposed towards the Community. Six-and-a-half years later, public opinion, particularly among Social Democratic supporters, was distinctly more Eurosceptical. Such was the magnitude of the crisis in 1990, and the pace at which external circumstances were changing as communism collapsed, that the question of EC membership sprang to the forefront of Swedish politics remarkably quickly; Eurosceptics had no time to organise and mobilise before the party's leaders had announced Sweden's application to join. In 1995-97 opponents of EMU, though never reaching anything like the level of organisation they achieved in the 1994 referendum campaign, were clearly present in the highest ranks of the party. In addition, whereas in 1989-90 SAP had been attempting to bargain in parliament with bourgeois parties that were generally supportive of the

---

[29] *Dagens Nyheter* April 20th 1997.
[30] *Dagens Nyheter* April 24th 1997.
[31] *Dagens Nyheter* May 14th 1997

EC, in 1995-97 the Social Democrats' main collaborators, the Left and the Centre, were both strongly against EMU.

Yet just as these conditions changed in respect to the decisions by the Social Democratic leadership in 1990 and 1997, so they changed again thereafter, as we shall see in the final chapter.

8 Conclusions: Swedish Social Democracy in the European Union

In this study we have examined the problem of why a highly successful political party should have had so many internal problems with the issue of European integration in general, and, more specifically and acutely, with whether to support membership of the European Union. The case we have examined, that of the Swedish Social Democratic Party, is a puzzle for various reasons. From a European perspective, it is unusual to see a centre-left party in such difficulty over European policy. SAP's sister parties in the rest of Scandinavia and, to a lesser extent, in the British Isles and France have contained significant Eurosceptical elements, but now Britain and Ireland's Labour parties and the French Socialist Party broadly conform to the West European norm—that is, of mainstream social democratic parties being in favour of closer integration between the members of the EU. From a Swedish perspective, the disharmony within SAP is, if anything, even more puzzling. It can be justifiably stated that, in what is over a century of the party's existence, European integration has divided it like no other issue. Defence has sometimes caused disagreement among Social Democrats, and nuclear power caused considerable discord from the mid-1970s. But they did not divide SAP so squarely down the middle. In any case, the party's history is marked much more by unity and discipline. Without those qualities, it is unlikely that it would have spent just 14 years since 1921, and nine since 1936, out of office.

The unusualness of the division that SAP suffered over Europe raises two possibilities: first, that something about the party changed to make it more vulnerable to internal division; or, second, that there was something about the issue of European integration that was peculiarly divisive. The broad conclusion of this study is that, to differing degrees, both these possibilities apply. Something has changed and is changing in the political behaviour of parties and voters. But the EU is an issue that, in

181

certain circumstances, can divide some parties in some countries in a way that few other issues can.

The study's approach to the topic centred on two main research questions. First, why did SAP's leadership change its policy towards the EU so abruptly? And, second, why was the party so divided by the leadership's decision to do so? In summarising the results of previous chapters and drawing together their conclusions, we can offer answers to those questions. We can do so most effectively by reversing their order—that is, by starting with the general (the attitudes of Swedish Social Democracy to European integration) and moving to the particular (why the party changed its European policy in 1990).

Why was Swedish Social Democracy divided over Europe?

There is evidence that voters do approach European integration by seeking to weigh up rationally the costs and benefits involved in supporting or opposing it. Sometimes the grounds for this cost–benefit analysis are actually quite clear and tangible. Anecdotal evidence alone suggests that countries like Portugal and Ireland have been transformed, not least in their infrastructure, during their time in the EU. Voters must see these benefits, and incorporate them into their view of integration. Furthermore, there is no reason why we must limit our variables in this analysis to economic interest. Gabel and Palmer, for example, incorporate a security element in their attempt to explain variance in public support for integration.[1] Although their operationalisation of this hypothesis is debatable, their basic conception of attitudes to European integration is quite reasonable. It is possible to see individual political actors—voters and activists as much as politicians and bureaucrats—as approaching it on the basis of rational analysis of costs and benefits to each, according to the information they have.

As the result of the referendum on EU membership showed, Sweden is a comparatively Eurosceptical country, and to explain the division within SAP, we need to offer some suggestions of why that might be. The first of this study's major conclusions is that, *for Sweden, the costs and benefits of EU membership were finely balanced.* Unlike most member states, Sweden has no "big reason" for belonging to the EU. Regarding

---

[1] Matthew Gabel and Harvey D. Palmer, "Understanding Variation in Public Support for European Integration", *European Journal of Political Research* vol. 27, 1995.

trade relations, Sweden was already a highly integrated member of the West European economy; full EU membership did not confer many more advantages in terms of market access, nor in incorporation in EU policy regimes. Conversely, had Sweden not joined the Union, the EEA would have preserved most of the advantages the country did have. Concerning security, EU membership did not constitute a "soft" security enhancement in the way that it did for Finland vis-à-vis Russia—or, for that matter, for the Low Countries or France vis-à-vis Germany. Nor would saying No to full membership have diminished Sweden's security, given its wish to retain nonaligned status in any case.

Thus, Sweden offers a case in which it is hard to apply assumptions of voters' rationality to their views of European integration. Rational voters simply lack much clear and tangible evidence of whether their self-interest would be better pursued through support or opposition to EU membership. That was the general finding of this study's questionnaire survey. Although nearly half of those Social Democratic activists planning to vote against membership thought their own economic circumstances would be worsened by EU membership, the majority of all categories of respondent—that is, those intending to vote Yes, No or who remained undecided—were all uncertain whether EEA or EU membership would have any effect on their own economic circumstances (see table 8.1).

When it is unclear how to pursue self-interest, voters are likely to look for cognitive "short-cuts"; other variables, not obviously based on rational pursuit of self-interest, will then become more relevant. Factors such as national culture and political ideology are thus very germane to explaining attitudes towards the EU. They may be second-order variables, liable to supersession by first-order ones, such as material and economic advantage, or military security. But in the absence of such overriding factors, ideology and culture can play a significant role in shaping preferences. It is the second major conclusion of this study that *the ideology of Swedish Social Democracy is less compatible with the concept of supranational integration than most brands of European social democracy.*

Chapters 3 and 4 presented the evidence for this assertion. From an early stage, Swedish Social Democracy became reconciled to living with a market economy. Free trade with the rest of the world was embraced; at home, private business was generally allowed scope to operate. Certainly, the Social Democrats' domestic economic policies did not amount to free-market liberalism. The high taxes and wide-ranging, universal state-

Table 8.1 Social Democratic activists' views of their own circumstances and EEA and EU membership

Referendum voting intention

	Yes %	Uncertain %	No %	total %

How do you think your personal economic circumstances will be affected by Sweden's EEA membership?

	Yes %	Uncertain %	No %	total %
Improved	20.3	11.8	11.8	15.2
Uncertain	77.1	76.5	71.8	75.0
Worsened	2.5	11.8	16.4	9.8
total	100	100	100	100
	(N = 118)	(N = 68)	(N = 110)	(N = 296)

How do you think your personal economic circumstances would be affected by EU membership?

	Yes %	Uncertain %	No %	total %
Improved	22.9	4.4	1.8	10.7
Uncertain	72.9	70.6	52.7	64.8
Worsened	4.2	25.0	45.5	24.5
total	100	100	100	100
	(N = 118)	(N = 68)	(N = 112)	(N = 298)

welfare schemes that they introduced could scarcely be described as such. The labour market was strongly influenced by state intervention, directly in the expansion of public-sector employment and through the Labour Market Board, and indirectly in the form of wage solidarity. In its radical phase in the late 1960s and 1970s, SAP sought to go considerably further, and presented the wage-earner funds as a means of wresting control and even ownership of business away from private shareholders. But there is evidence that this was an aberration, and that SAP has usually had a more tolerant view of markets than most other social democratic and socialist parties. There has always been a significant element within the party that has seen the market, not as an end in itself, but as an invaluable tool for achieving Social Democratic goals, such as prosperity and equality.

This makes Swedish Social Democracy distinct. So does another of its ideological characteristics: the appropriation of national sentiment and symbolism for its own purposes. In European countries, it has usually been the political right that has managed to secure the mantle of "party of the nation". Social democracy was borne out of working-class opposition to ruling elites. As the elites held the organs of power in the nation states, it became natural for social democracy to seek to bypass those organs, and stress international solidarity between oppressed proletariats. In the Protestant countries especially, such as firmly Lutheran Sweden, nationalism became a potentially powerful political weapon. What made Sweden a special case is that, for various reasons (the timing and pace of industrial revolution, the peaceful realisation of universal democracy, the consequent strength of the labour movement), it was the left that managed to grasp and exploit this weapon, not the right. The nationalism of Swedish Social Democracy developed into a welfare nationalism, evidence of which we saw so clearly in the book by Ekström, Myrdal and Pålsson.

Thus, the likely suspicion of European integration among Social Democrats can clearly be understood. Our observations received general, if not unqualified, support in our questionnaire survey of party activists, which we presented in chapter 4. Far from seeing the nation as a source of oppression and war, as many social democrats on the continent do (for obvious historical reasons), for SAP the nation epitomised all the party's successes. Sweden's democratisation had been early, peaceful and durable, not violent or prone to extremist usurpation; indeed, the principles of democracy and national sovereignty became associated in a way that (also for obvious reasons) was incomprehensible in other European countries. While other nations had been ravaged by conflict at least twice during the 20th century, Sweden had remained intact and free of war.

Perhaps most importantly, the welfare state was based upon this sense of nationhood. The welfare state can be seen as a form of mutual insurance; it depends on a sense of empathy between contributors and recipients. Shared nationality is what ultimately underscores all European welfare states. Sweden had that shared nationality, and upon it its Social Democrats proceeded to build a universal welfare system of which they were immensely proud. So, many of them naturally asked, in the 1960s and 1970s especially, what could Sweden gain from economic and political union with countries without such attributes, without such achievements? The EC/EU could not even give Sweden many advantages in terms of market access, so integrated was the country already in the wider West European economy. In fact, the organisation's free-trading credentials were rather dubious; to many Social Democrats, it looked primitively protectionist and mercantilist. All the time, and quite naturally, SAP's leaders sought to exploit this sense of national pride for electoral purposes by actively emphasising the superiority of the Swedish model—and the party's credit for it.

This historical and ideological context, then, forms the backdrop to our study of party behaviour. It is a constant in our analysis. Together with it not being obvious what in economic and security terms Sweden has to gain from EU membership, it explains why Euroscepticism among Social Democratic activists, members and supporters should have been so strong. With these factors in mind, it also becomes easier to understand why post-materialist elements in the party—which appear to have become significant, even if the psephological evidence is as yet less than convincing—should have been inclined towards looking sceptically at the EU's remote, bureaucratic and materialist character, rather than seeing it as a platform for cosmopolitan and progressive politics, as is more the case in some European countries. The question now, of course, is why, given all this, the party leadership discarded its own Euroscepticism so spectacularly in 1990 and thereafter. This brings us to the third major conclusion to this study. It is that, *because of rapid changes in the domestic political environment in the years up to 1990, the means necessary for individuals at different levels of the Social Democratic Party to pursue their goals diverged sharply.* These environmental changes concerned in particular the institutional setting of Swedish party politics, and the advent of economic crisis.

Institutional change

Usually, political parties find a balance between the interests of their elites, grass-roots and other constituent organisations. Elites may want the power and material rewards of holding office, but they cannot be so opportunistic in pursuing them as to alienate their more idealistic followers. The grass-roots, meanwhile, may want to see their preferred policies implemented, but they know that this will be impossible unless their party leaders win sufficient votes to propel them into office, or at least to wield influence in parliament by capturing pivotal seats. In social democratic parties, trade unions may want to promote their narrow sectional interests, but they know that doing so too blatantly could cost the party votes, which might retard their pursuit of their goals. SAP's electoral history suggests that it has usually managed to balance these interests successfully. But analysis of party organisation broadly supports the hypothesis that it has increasingly been the leadership's interests that are dominant in shaping party strategy. The leadership has a strong position in the party's power structure, predicated not least on its access to public funding, and its control of a large and professional party bureaucracy.

As for the means of promoting those goals, we also saw in later chapters a little of how the Swedish political environment has changed in recent years. Various scholars agree that the consensual, hegemonic nature of Sweden's politics developed into something more conflictive and bloc-based after the 1970s.[2] This was evident both in government formation and in the everyday work of the Riksdag. In 1976 the non-socialist parties secured a parliamentary majority for the first time in the democratic era, and retained it (by the narrowest of margins) in the election of 1979. Although the Social Democrats recovered in 1982 to govern throughout the rest of the decade, the non-socialists formed another administration, albeit a minority one, in 1991. At last there seemed a viable governing alternative in Sweden that excluded the Social Democrats. In addition, the rise of new parties further upset the status quo ante. In 1985 the Christian Democrats became the first new party in the Riksdag for 64 years. In 1988, the Greens won 21 seats in the Riksdag. A completely new party, the right-wing populist New Democracy, also entered parliament in 1991. Moreover, as the Communist Left metamorphosed into the Left Party, it became a less reliable ally for SAP.

The polarisation of politics can be gauged by the change in party relations in the Riksdag's 16 standing committees. For many years, unanimity in the committees' reports was almost a tradition; between the 1930s and the mid-1950s around eight in ten were concluded in this way. But agreement between the four major parties (the Social Democrats, Centre, Liberals and Moderates) became less frequent by nearly a fifth during the 1980s, and disagreement in committee reports that conformed to bloc positions (that is, SAP against the other three), which had occurred in just 20 per cent of cases in 1972, broke out in 35 per cent in 1983-84 and in over 40 per cent in 1986-87.[3]

These changes in the party system, combined with certain institutional reforms—unicameralism, a more proportional electoral system and

---

[2] Cf. Einhorn and Logue (1988); Olof Ruin, "Sweden: The New Constitution (1974) and the Tradition of Constitutional Politics", in Vernon Bogdanor (ed.), *Constitutions in Democratic Politics* (Aldershot. PSI/Gower. 1988); Sannerstedt and Sjölin (1992).

[3] Sannerstedt and Sjölin (1992), p.113, 127.

negative parliamentarism[4]—all conspired to make the battle for votes more intense in Swedish party politics. It is very possible that the heavier the emphasis on vote-maximisation in a party's strategy, the greater the pressure will be on its internal balance between the leaders' goals of attaining office and the grass-roots' goals of policy implementation. An electoral platform is a compromise between the desires of various lobbies, and the more voters that it seeks to attract, the wider its compromises have to become. In other words, the more moderate and all-embracing—or, to use Kirchheimer's phrase, catch-all—the platform must be. Unless they are motivated by selective (material) benefits (which seems rarely to be the case in Sweden), people must be motivated to join and/or become active in a party by the hope of seeing it implement certain policies. Members and activists are thus likely to resent these policies being ever more diluted by the leadership in the pursuit of votes. Just such a process seems to have been ongoing in SAP in the late 1980s and early 1990s. An activist interviewed in Örnsköldsvik, in northern Sweden, felt that the attempt to impose a ban on strikes during the war of the roses in 1989, and the brief appearance of New Democracy in 1991-94, were both symptomatic of established parties losing touch with their members and voters. It appears also to be happening more widely in Western Europe.[5] But for SAP, the issue of EU membership turned a gradual process into an immediate and acute one.

The crisis of political economy

These changes in what we might call the sociology of a political party surely contributed to the difficulty Social Democratic leaders experienced in the 1990s in persuading their grass-roots of the merits of European inte-

---

[4] Negative parliamentarism signifies that the formation of government did not require a majority vote to *approve* it, while a vote of no-confidence to *remove* a government did. Bergman argues persuasively that this was designed deliberately—by the Social Democrats in particular—to make minority governments more likely and majority coalitions less likely. Torbjörn Bergman, *Constitutional Rules and Party Goals in Coalition Formation: An Analysis of Winning Minority Governments in Sweden* (Umeå, Department of Political Science, Umeå University, 1995), ch. 7-8.

[5] Lane and Ersson, for example, go so far as to question whether the old assumptions of Lipset and Rokkan's social cleavages are even very helpful now, such has been the rise electoral volatility. Jan-Erik Lane and Svante Ersson, "Parties and Voters: What Creates the Ties?", *Scandinavian Political Studies* vol. 20, no. 2, 1997.

gration. Yet there was a more fundamental reason why the EU should have become such an acutely divisive issue for SAP. It was that the volte-face over Community membership was intimately connected to the party's failure, in terms of practical policy and intellectual strategising, to address the growing crisis in the Swedish economy.

Joining the EU did not accord with pursuing the membership's goals: it clashed with long-standing assumptions and prejudices, and there was no major reason, related either to economic or security needs, that was sufficient to persuade the grass-roots that those assumptions and prejudices should suddenly be discarded. By contrast, membership quickly came to be seen by the party's elites as being imperative for the pursuit of their goals. As the political and economic situation deteriorated in 1988-90, Social Democratic elites became increasingly unable to deal with the crisis. As the party's opinion poll ratings fell, so the leadership's main strategy—vote-maximisation—seemed to be falling apart. The Liberals and Moderates used the European issue skilfully as a weapon to attack the government, and there was some sense of SAP shifting its position in tandem with a change in public opinion. But far more important was a desire, not so much to catch more votes by changing policy, but to forestall a really catastrophic further loss of public support that might have come with economic meltdown, and a forced devaluation of the krona in particular. This is what chapter 5 described.

The decision to reverse the party's European policy in 1990 was, then, a truly critical juncture. For many rank-and-file Social Democrats, it came to represent the abandonment of the traditional goals and strategy of the Swedish labour movement, full employment above all. This a perception was probably sharpened in retrospect by the great rapidity of the change. At the time, the crisis also provided a golden political opportunity—not just for the bourgeois opposition, but also for the modernising elements in SAP (the end of the third way notwithstanding). There is scope for disagreement, even before the wage-earner funds were scaled down and Sweden's financial markets were deregulated in the mid-1980s, as to whether there was any real chance of reverting to the economic policies that became popular in SAP during the 1960s and 1970s. It is clear, though, that for those Social Democrats who felt that such a return was impossible, undesirable or both, the EC offered the *means* to encourage, among other things, a non-accommodating macroeconomic policy and a more streamlined public sector. Moreover, the crisis of 1990 offered the *opportunity* for these modernisers to secure this policy goal. There is, therefore, as chapter 5 illustrated, a significant element of left against right, moderniser against traditionalist, in the Social Democratic division over Europe, even if it is more complex than that.

Of course, the theories used here cannot explain—let alone predict—why one Social Democrat, influenced by his party's ideological history, should have voted No to EU membership, while the next, with whom the first agrees on all other political issues but who is ultimately swayed by loyalty to his party leadership, should have voted Yes. That is a question for psychology rather than political science. But presented here are the long- and short-term factors that made it such a difficult decision for Social Democrats, and which help explain why so many members and voters were prepared to reject the party leadership's advice.

Implications of the study

It was suggested at the start of the chapter that, in investigating the problem of SAP's division over European integration, this study has identified forces that can make a party more vulnerable to disunity than it has been hitherto, but also that there is something peculiarly divisive about European integration. These points can now be outlined briefly. But again we start this section at a still more fundamental level: assessment of our theoretical assumptions and methodological tools.

Theoretical and methodological implications

This study cannot be said to have tested a particular theory, or even a group of theories, in the true sense of the word. What it has done is (a) to start with the research problem, (b) to examine theories of party behaviour that might be useful in explaining our research problem and (c) to apply them throughout the preceding chapters. Finally, we offer our conclusion: that SAP's disunity can be understood as ultimately a conflict of interests among rational, self-interested individuals within the party, albeit with their rationality "bounded" by the their cultural, historical and ideological context, and with their self-interest in this issue often not easy to identify.[6]

---

6 Cf. Herbert A Simon, "Human Nature in Politics: The Dialogue of Psychology with Political Science", *American Political Science Review* vol. 79. no. 2, 1985.

This explanation can now stand against other possible explanations (based perhaps on theories of inexorable economic necessity, or political culture). It can also be tested, more scientifically, in other cases of party division over European policy. Possible candidates from Britain and elsewhere in the Nordic region spring immediately to mind, but we may see equally good ones emerging in post-communist Europe, as EU membership becomes more likely for several countries.

This has been a study of party behaviour. But there is scope for using its approach, with just a little adjustment, to investigate the broader problem of attitudes to European integration. Attitudes to national governance are usually analysed by using electoral data as the dependent variable. This is much harder when European-level governance is the topic. Its remoteness and abstract nature mean that electoral tests of public opinion towards it (referendums and elections to the European Parliament) are poor indicators. It is usually difficult for voters to see much connection between their choice in the ballot box and measurable political difference; no European government is drawn from the European Parliament, for example. Thus, such tests of opinion are inevitably "polluted" by national political issues.[7] Moreover, the vague and hypothetical questions used in surveys like the EU's Eurobarometer, in which it is especially hard for respondents to associate concrete consequences with answers, make this type of survey data dubious. But the study of parties offers a promising way of resolving this problem—in effect, of providing the researcher with a dependent variable. For all the changes in their role that scholars have suggested, parties do still retain a function as mechanisms for linkage between governors and governed. Because vote-seeking is part of parties' intrinsic purpose, they must tailor their European policies to attract votes. Their European policies can thus be used as indicators of public attitudes to integration. Focused comparative analysis that takes account of and controls for certain vari-

---

[7] For example, it is often hard to separate national and European causes of voting behaviour. The rational voter may ask whether, for example, his life will be measurably better or worse if, on one hand, a European treaty is carried or falls, or, on the other, he rebuffs a government he does not like by voting against its recommendation that he approve the treaty. Few observers inferred that the French electorate was badly divided over European integration because the country's referendum on the Maastricht treaty in September 1992 approved it so narrowly (51 per cent to 49 per cent). Rather, it was considered keen to punish an unpopular president for calling the poll at all. Cf. Gerald Schneider and Patricia A. Weitsman, "The Punishment Trap: Integration Referendums as Popularity Contests", *Comparative Political Studies* vol. 28, no. 4, 1996.

ables—geopolitical circumstances, economic situation, ideological tradition, internal party organisation, national rules of the political game—can be invaluable in investigating the fascinating political question of why, for example, Swedes and Germans, who in most respects have much in common, take such apparently different views of the desirability of further supranational integration.

Substantive implications

It is surely not surprising that throughout the European Union, political elites tend to be markedly more sympathetic towards integration than their parties' rank-and-file. Vaubel, for instance, offers a public-choice perspective on the centralising dynamic contained in the EU's institutional structure, and identifies various ways in which it can be helpful to national elites. The Council of Ministers he calls a "cartel of politicians", in which governments can co-operate in delegating to the EU the work of satisfying domestic lobbies, when to do so at home would be controversial. The Union's model of "executive federalism" can be exploited to bypass the scrutiny and control of national parliaments. Moreover, the promise of highly paid jobs, especially in the Commission (58 of the 82 commissioners since 1958 have previously been national politicians or bureaucrats), can encourage representatives to favour more integration and centralisation of power.[8] In short, the EU offers new opportunities for political elites to fulfil their office-seeking goals. But for the parties' grass-roots, the opportunities are much fewer. In countries like Sweden, where there is no general consensus about the benefits or otherwise that the country derives from EU membership, this divergence of views and interests could be a serious bone of contention within parties.

The impact of the EU on European parties must be seen within the context of broader changes in those parties. There does seem to be a trend towards the marginalisation of the membership and activists. The development of cheap and efficient methods of mass communication has made the membership less indispensable for disseminating parties' messages to voters. Within the party, this also applies to communication between leaders and members, which has had the effect of making intermediate party infrastructure—local bodies, federal components and the like—similarly

---

[8] Roland Vaubel, *The Centralisation of Western Europe* (London, Institute of Economic Affairs, 1995), pp.41-43.

marginalised. In its most extreme form, these trends may be making parties more like business firms, with the leader–voter relationship approximating more closely the owner–customer one.[9] Perhaps a more accurate analogy can be offered if the party-as-business-firm is understood as having the state as a major shareholder. Katz and Mair, for example, propose the concept of the "cartel party", one that forms with its rivals a group of quasi-competitive policy platforms, and which, largely because of their reliance on public funding, reach out directly to the voter from the state.[10]

Normatively, it is unclear whether this development is desirable. It is possible to argue that a more direct relationship between politician and voter, and the concomitant marginalisation of ideologically driven activists from the political process, are no bad things. However, in countries and parties where the tradition—or at least a perceived tradition—of grass-roots, participatory democracy is strong, it is viewed with alarm. SAP is an excellent example of such a party. In so encompassing a movement as that of Swedish labour, a drift towards pragmatic, non-ideological politicians, and coldly professional party bureaucrats, as well as passive, disinterested voters, is sometimes seen as a danger to the culture of *folkstyre*.

There may be something in such fears. But when we reconsider the phenomenon of European integration, concern is amplified. In the national setting, even if political elites are escaping the constraint imposed by their parties, they are at least still bound by their basic need to retain the approval of the national electorate. In the Union, meanwhile, much is made of the fact that some key decision-makers, such as commissioners and judges, are unelected. But rather more serious may be Vaubel's critique of the most powerful decision-making body, the Council of Ministers. Each of its members is accountable to only a fraction of the voters that the Council's collective decisions affect. Each can thus promote the narrow sectional interest of his country, or more often a still narrower economic interest within that country, without considering the wider interest of all the EU's citizens. Such narrow interests—farmers are the best example—can be indulged with the provision of public goods (subsidy, trade protection, market regulation) at the European level. This is costly to the Union

---

[9] Cf. Jonathan Hopkin, "New Parties and the Business Firm Model of Party Organisation: Cases from Spain and Italy", paper presented to Political Studies Association annual conference, Glasgow University, 1996.

[10] Cf. Mair (1994); Richard S. Katz and Peter Mair, "Changing Models of Party Organization and Party Democracy: The Emergence of the Cartel Party", *Party Politics* vol. 1, no. 1, 1995.

as a whole, but its voters lack the means of expressing a collective interest that might offset the influence of (to use the economic term) rent-seeking groups.

The commonly suggested remedy for this lack of democratic accountability is to promote the power of the European Parliament. But this would scarcely correct the problem. Members of the European Parliament (MEPs) campaign on, are elected on and, most importantly, are held accountable on the basis of national issues. Far from offsetting national sectional interests, MEPs represent them. Only when genuinely European parties can form, offering EU-wide policy platforms to a European electorate, and even aspiring to identify and capture a European median voter, will the collective interest of European citizens be represented in the European Parliament. Such is the enduring strength of national identity—the "national cleavage" between Europeans, to adapt Lipset and Rokkan's term—that such parties look inconceivable in the foreseeable future. As we have seen, national identity is especially strong in Sweden.

Our introductory chapter suggested that European social democratic parties are facing a choice between the demands of politics and of democracy; that is, between maximising their influence over the market through EU-level public institutions, and keeping the individuals who administer them subject to effective democratic accountability.[11] This is a normative, ideological question. But most (if not all) politics is still about the distribution of economic resources. Ideology takes second place when EU membership offers a country clear and tangible economic (or security) benefits. In Sweden, where such benefits are absent or doubtful, the Social Democratic Party has been badly damaged by disagreement over whether politics or democracy should take precedence. If and when the Union expands further, some existing member states may find the benefits they currently derive from membership diminishing. Their social democratic parties may then find their grass-roots and voters becoming similarly unhappy with their leaders' priorities.

Swedish Social Democracy in the European Union

The Social Democrats endured a difficult period in office after their election victory in September 1994. As we saw earlier, tax rises and spending

---

11 The Calmfors report made a similar point. SOU 1996: 158. p.424.

cuts were necessary, and these inevitably caused pain and controversy. Although the austerity policy was more effective in repairing the public accounts than nearly anyone expected, the government's unpopularity inevitably suffered. In February 1997 SAP briefly fell behind the Moderates in the polls for the first time ever, although the conservative advance petered out thereafter. At the same time, relations with the private sector became very strained. Over 100 business leaders complained about the government's failure to promote a favourable climate for business, especially in light of the decision to decommission the nuclear power stations.[12] Ericsson, Sweden's biggest export firm, decided to move part of its headquarters to London, claiming that income-tax levels made it unattractive for managers to live in Sweden. After Carlsson announced his retirement, and Sahlin had become embroiled in the credit-card scandal, there followed a highly embarrassing interlude for the Social Democrats, during which none their other potential candidates for leader seemed to want to stand. The debacle reinforced the perception that leading the modern Social Democratic Party had become a rather thankless task, and Persson's eventual acceptance of the job, after repeatedly denying any interest in it, was scarcely the most auspicious start to a premiership.

But even as these episodes passed, the party's biggest difficulty arose from the Persson government's failure, despite the prime minister's reckless early promises, to halve unemployment by 2000. After Persson's first year in the job, the level had actually risen, standing at around 8 per cent—which would be some 4-5 per cent higher if those on government make-work and training schemes were included.

---

[12] Tommy Adamsson *et al,* "'Nu är förtroendet förstört'", *Dagens Nyheter* February 23rd 1997.

Figure 8.1 Swedish public opinion and EU membership

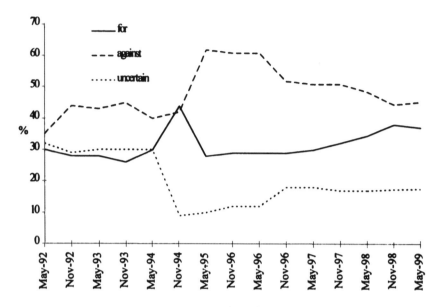

Source: Statistics Sweden.

Europe, then, was by no means the Social Democrats' biggest problem; yet it remained a headache for the party leadership. Although the issue was becoming a question of economics and politics, rather than one of security policy,[13] EU membership remained generally unpopular in Sweden (see figure 8.1), and this naturally complicated the government's policy-making. The obligations of membership—from budget contributions to the CAP to a foreign court deciding Swedish alcohol policy[14]—were not always welcome. Nor were the benefits of membership apparent to many Swedes. Their economy did not improve noticeably after accession. Cuts in public spending, which the government had implied would be the consequence of a No in the referendum of 1994, were

---

[13] Cf. Kite (1996), who detected this crucial difference in the parliamentary debates on EC membership in the early 1990s compared to those that had taken place in the 1970s and 1980s.

[14] Despite an advocate-general's preliminary ruling the previous March, in October 1997 the European Court of Justice ruled that Sweden's state monopoly of alcohol retail was in fact compatible with European law.

implemented anyway. Even specific advantages that some supporters of membership had promised failed to materialise: food prices, which had fallen by around a third in Finland after it joined, hardly changed in Sweden.

There have undoubtedly been successes for the government in European matters. The intergovernmental conference of 1996 could have been an immediate source of difficulty for the Social Democrats, but European leaders' appetite for further integration, which had culminated at Maastricht, had waned, and the Treaty of Amsterdam, signed in June 1997, was fairly uncontroversial in the party. Indeed, the government's initiatives in the pre-Amsterdam negotiations had some success. Clauses designed to increase the transparency and openness of EU administration were added to the treaty. The "Petersberg tasks" of peacekeeping were given specific mention in the common foreign and security policy, without establishing a closer relationship between the EU and NATO or the WEU. Perhaps most important symbolically, the Swedish proposal for an "employment chapter" found its way into the treaty. This last element clearly illustrated the attempts by the Social Democratic leadership to show its party that the EU level could be valuable in pursuing its traditional goals, and was not just a force for macroeconomic stringency. Whether anything substantive will flow from the employment chapter, or from the provision for more bureaucratic transparency, must be questionable, however. In fact, arguably the biggest success for the Social Democratic government at Amsterdam was what the treaty left out. Potentially far-reaching institutional changes, for instance—which France and Germany had been mooting in advance of further enlargement, and which Sweden's chief negotiator, like representatives from other small member states, had publicly declared unacceptable—were dropped.

Another success for Sweden was the decision, in December 1997, by the European Council that the Union would not focus on just a few candidates for membership from the ranks of the former communist countries, but rather would treat all the applicants—ostensibly, at least—on an equal basis. With the three Nordic members particularly keen to encourage the accession of the Baltic states to the Union, this decision was considered a triumph for the Swedish position.

Yet EMU continued to overshadow all other progress that the Social Democratic government could claim in Europe. The announcement that Sweden would not be participating in 1999 drew a few critical comments from elsewhere in the EU—Luxembourg's prime minister described it as "rich", for example[15]—but overall the argument that it would be in no one's interest for the country to adopt the euro without the necessary domestic political support was generally acknowledged.

In any case, when in March 1998 the European Commission announced which member states it was recommending for participation in EMU's stage 3, Sweden was rejected on the grounds that the krona had not participated in the ERM for the previous two years, and that the Riksbank did not meet the Maastricht treaty's stipulation that national central banks be independent of political control. Although the Swedish government accepted the decision, it also sought to keep its options open for future accession to EMU. It declared that the Riksbank's legal position was being made compatible with the requirements of the EMU system: an agreement between five Swedish parties to increase the Riksbank's independence in policy-making was announced in May 1997, and came into force at the beginning of 1999. While the krona had never been part of even the post-1993 ERM, the government argued that it had actually been more stable against the D-mark since than some other currencies that had been members. The government insisted further that the remaining aspects of its monetary policy were compatible with the Stability Pact, and also that Swedish firms would be permitted to account in euros. A high-level steering group, chaired by Erik Åsbrink, the finance minister, was announced to co-ordinate policy towards EMU.[16]

The governor of the Riksbank made no secret of his preference for Swedish participation in EMU as soon as possible. Åsbrink's sympathy with this view was also fairly clear. Persson's real intentions and preferences vis-à-vis monetary union remained opaque, however. In October 1997 he again made Eurosceptical noises in a magazine interview. "I don't want a federal development [in Europe]," he declared.

> The Swedish people have not accepted a federation. EMU can certainly lead to demands for a European economic policy. That would not be possible to develop, if it were powerful, without also building up common European political organs. The federal element would grow from that. This discussion is central in the context of EMU.

---

15 *Svenska Dagbladet* March 23rd 1998.
16 Ministry of Finance, "EMU Facts", 1998 (www.regeringen.se).

Federalism, he argued, "would lead to a tremendous distance, a chasm, between decision-makers and voters. I don't believe that would be seen positively in Sweden." It would be possible to judge in 2001-2 at the earliest whether the EU was becoming a federation, Persson said; but he was less optimistic about this than he had been a year before.[17] Yet the prime minister insisted that he was "genuinely uncertain" whether monetary union was desirable for Sweden.[18] Meanwhile, he oversaw the launch in autumn 1998 of a government information campaign about EMU, for which a fifth of the SKr19bn cost was, to the dismay of Social Democratic EU-Critics, provided by the European Commission.[19] Such an information campaign, of course, was precisely what had preceded the Social Democratic leadership's push for acceptance of EU membership in 1991-94.

Whether the election of September 1998 brought Sweden closer to adopting the euro was unclear. The parliamentary debate of December 1997 had marked a change in the Moderates' strategy, as Bildt dropped the demand that Sweden participate in 1999, and accepted that a referendum, perhaps in tandem with the 1999 European elections, would be necessary before Sweden did join. In effect, this ended the issue's salience in the election—which, given the Social Democrats' divisions, suited the party well. Remarkably, the Social Democratic manifesto did not even mention monetary union.

At first glance, the outcome of the election diminished the chances of an early Social Democratic commitment to EMU. Persson continued as prime minister, but his party's disastrous score of 36.4 per cent, its worst since universal suffrage was achieved in Sweden in 1921, left it needing not one but two allies in the Riksdag. Persson opted not to make overtures to the parties of the centre, and instead chose to rely for parliamentary majorities on Sweden's two most Eurosceptical parties, the Left—which won 12 per cent, its best result ever—and the Greens. The election's other big winners, the Christian Democrats, whose 11.8 per cent was also its highest-ever score, had opposed Swedish accession to EMU in 1999. Of the two parties that were keenest on early entry, the Moderates were disappointed to see their vote rise only marginally, despite the government's unpopularity, while the Liberals lost support again, and looked to be drifting perilously towards the threshold for representation in the Riksdag.

---

[17] *Veckans Affärer* October 20th 1997.
[18] *Aktuellt i Politiken* November 27th 1998.
[19] *Finanstidningen* October 13th 1998, October 27th 1998.

Moreover, as SAP licked its wounds after its drubbing, the notion of broaching a controversial, divisive question such as the euro was not immediately attractive.

Yet several developments in 1998 probably made Swedish accession to EMU early in the 21st century more likely. The social democratic advance in the EU's member states, which had seen centre–left governments win power in Britain and France the previous year, was continued with the election of an SPD-led government in Germany in September. For the first time, the Union's three biggest countries had social democratic governments at the same time, and Germany's in particular began to talk of co-ordinating a common macroeconomic strategy against unemployment as a complement to EMU. The British government, though appearing less enthusiastic about such plans, did commit itself to the principle of EMU and to joining at an appropriate moment. Moreover, in the run-up to Sweden's own election in 1998, currency instability flowing from the economic crises in Asia and Russia buffeted the Scandinavian currencies— including the Danish krona, despite its tried and tested peg in the ERM— while leaving the Finnish markka unscathed. Some observers attributed this to Finland's impending accession to monetary union.[20] Opinion polls suggested that the Danish electorate was warming towards EMU as its launch date drew closer.

Moreover, the parliamentary situation vis-à-vis EMU was possibly less clear-cut than it seemed. Leading Social Democrats made it clear after the election that European policy was off-limits for its new partners in the Riksdag. Any future decision by the Social Democratic leadership to support Swedish accession to EMU would very probably win the support and co-operation of the Liberals and Moderates. It might conceivably attract the other two non-socialist parties as well. In April 1998 Olof Johansson announced his retirement as Centre leader. His successor, Lennart Daléus, sounded more open to rapprochement with the other non-socialist parties; he had also reserved his position when the Centre's National Executive voted against EMU nearly two years previously. By late 1998 even elements in the Greens and the Left had begun to urge their leaders to take less Eurosceptical positions.

20 *The Economist* September 5th 1998.

Figure 8.2 Social Democratic supporters and EMU

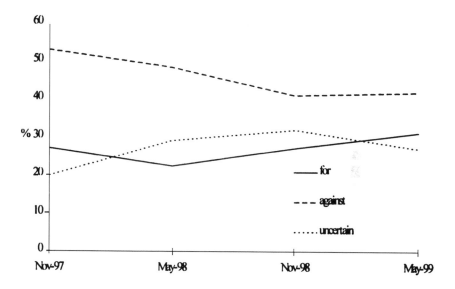

Source: Statistics Sweden.

Polls suggested that Social Democratic voters were gradually becoming less suspicious of EMU (see figure 8.2). In a survey of the party's parliamentary group, although a majority remained uncommitted, half of the Social Democrats' MPs either supported participation in EMU or leaned towards that option. Just over a tenth either opposed EMU or inclined towards that view.[21] Two MPs announced plans "to start a network for a positive debate on EMU in the Social Democratic Riksdag group and other parts of the party".[22] Persson's creation after the election of a new, high-powered Economics Ministry under two "super-ministers", Björn Rosengren and Mona Sahlin (whose appointment marked her political rehabilitation), added weight to the pro-EMU element in the Social Democratic cabinet. This, plus the prime minister's bringing responsibility for coordinating European policy away from the Foreign Ministry and into the cabinet office, closer to his personal control, were widely seen as making easier a commitment by the government to the single currency.

---

[21] *Aktuellt i Politiken* November 24th 1998.
[22] Reynoldh Furustrand and Pär Axel Sahlberg. "'Regeringen är för defensiv'", *Dagens Nyheter* October 21st 1998.

Nevertheless, it will probably be external developments—principally, the relative economic performances of Sweden and the euro-zone countries, and Britain and Denmark's decisions on whether to join EMU—that will be the crucial factors, and it will probably be the debate within the Swedish labour movement that decides the Social Democrats' approach to EMU in the coming years. There remained a significant number of sceptics about EMU in the cabinet, and there remained frequent public criticisms of the government's European line from Eurosceptics in the party. The leadership's position was still very finely balanced. Unless external events tilt the party's internal debate on monetary union in a particular direction, the decision on whether or not to support Swedish participation could be a tremendously difficult one. Unfortunately for SAP, its own short experience of life in the EU, plus that of some other social democratic parties in longer-standing member states, suggest that even when the question of EMU is decided, the next divisive development in European integration will not be too far behind.

Bibliography

Note: The bibliography and the index are ordered according to the expectation of the English-speaking reader, which means that Nordic letters are counted as their closest English equivalents. Thus, letters like å, ä and ö, which come at the end of the Swedish alphabet, are placed as if they were a, a and o respectively.

Adamsson, Tommy *et al*, "'Nu är förtroendet förstört'", *Dagens Nyheter* February 23rd 1997.

Adelsohn, Ulf (1990), "Carlssons motiv: en lugn kongress", *Dagens Nyheter* June 10th.

af Malmborg, Mikael (1996), *Den standaktiga national staten. Sverige och den västeuropeiska integrationen, 1945-59* (Lund, Lund University Press).

Allen, Hillary (1979), *Norway and Europe in the 1970s* (Oslo, Universitets-forlaget).

Åmark, Klas (1992), "Social Democracy and the Trade Union Movement: Solidarity and the Politics of Self-Interest", in Klaus Misgeld, Karl Molin and Klas Åmark (eds), *Creating Social Democracy: A Century of the Social Democratic Labor Party in Sweden* (Pennsylvania, Pennsylvania Universiy Press).

Andersson, Dan and P.-O. Edin (1997), "'Säg nej till EG, Carlsson!'", *Dagens Nyheter* June 24th.

Andersson, Ove (1994), "En rekommendation", *Aktuellt i Politiken* June 23rd.

Andrén, Nils (1991), "On the Meaning and Uses of Neutrality", *Cooperation and Conflict* vol. 26, pp.67-83.

Antola, Esko (1987), "EFTA-EC Relations After the White Paper", *EFTA Bulletin* vol. 28, no. 3.

Arter, David (1995), "The EU Referendum in Finland on 16 October 1994: A Vote for the West, not for Maastricht", *Journal of Common Market Studies* vol. 33, no. 3, pp.361-87.

Åsbrink, Erik (1996), "'EMU-anslutning kan senareläggas'", *Dagens Nyheter* August 28th.

— (1997), "Sverige och EMU—Anförande av finansminister Erik Åsbrink den 3 juni 1997 kl 18.30", Swedish Ministry of Finance, June 3rd (www.regeringen.se).

Åström, Sverker (1990), "'En onödig inskränkning'", *Dagens Nyheter* June 1st.

Aylott, Nicholas (1995a), "The European Union: Widening, Deepening and the Interests of a Small Member-State", EPRU Working Paper No. 3/95 (Department of Government, University of Manchester).

— (1995b), "Back to the Future: The 1994 Swedish Election", *Party Politics* vol. 1, no. 3, pp.419-29.

— (1997), "Between Europe and Unity: The Case of the Swedish Social Democrats", *West European Politics* vol. 20, no. 2, pp.119-36.

Bennulf, Martin (1992), "En grön dimension bland svenska väljare?", *Statsvetenskaplig Tidskrift* vol. 95, pp.329-58.

— and Per Hedberg (1996), "Passiv eller activ väljare?", in Mikael Gilljam and Sören Holmberg (eds), *Ett knappt ja till EU. Väljarna och folkomröstning 1994* (Stockholm, Norstedts juridik).

Berg, Jan O. (1994), "European Union Relations in Swedish Public Opinion: An Overview", in Centre for European Policy Studies, *The Fourth Enlargement: Public Opinion on Membership in the Nordic Candidate Countries* (Brussels, CEPS).

Bergman, Torbjörn (1995), *Constitutional Rules and Party Goals in Coalition Formation: An Analysis of Winning Minority Governments in Sweden* (Umeå, Department of Political Science, Umeå University).

Bergren, Henrik (1997), "Rätt diagnos, fel recept", *Dagens Nyheter* May 13th.

Bjørklund, Tor (1996), "The Three Nordic EU Referendums Concerning Membership in the EU", *Cooperation and Conflict* vol. 31, no. 1, pp.11-36.

Board, Joseph B. (1995), "Sweden: A Model Crisis", Current Sweden No. 410 (Stockholm, Swedish Institute).

Brodin, Katrina, Kjell Goldmann and Christian Lange (1968), "The Policy of Neutrality: Official Doctrines of Finland and Sweden", *Cooperation and Conflict* vol. 3, pp.19-51.

Calmfors, Lars, Harry Flam, Nils Gottfries, Rutger Lindahl, Janne Haaland Matlary, Ewa Rabinowicz, Anders Vredin and Christina Nordh Berntsson (1996), SOU 1996: 158 (the Calmfors report) (Stockholm, Ministry of Finance).

Carlsson, Ingvar (1990a), "'EG-medlemskap omöjliggörs'", *Dagens Nyheter* May 27th.

— (1990b), "'EG hinder kan undanröjas'", *Dagens Nyheter* July 5th.

— (1994), "Europasamarbetet: ett vänsterprojekt", in Rolf Edberg and Ranveig Jacobsen (eds), *På tröskeln till EU* (Stockholm, Tidens förlag).

— (1995), "'EMU för jobbens skull'", *Dagens Nyheter* September 10th.

Church, Clive (1991), "EFTA and the European Community", European Dossier 21 (London, PNL Press).

Cohen, Benjamin J. (1993), "The Triad and the Unholy Trinity: Problems of International Monetary Co-operation", reprinted in Jeffry A. Frieden and David A. Lake (eds), *International Political Economy: Perspectives on Global Power and Wealth*, 3rd ed. (London, Routledge, 1995).

Dagens Nyheter (1990), "Sverige närmare EG", June 12th.

Dohlman, Ebba (1989), *National Welfare and Economic Dependence: The Case of Sweden's Foreign Trade Policy* (Oxford, Clarendon Press).

Downs, Anthony (1957), *An Economic Theory of Democracy* (New York, Harper and Row).

Dyson, Kenneth (1994), *Elusive Union: The Process of Economic and Monetary Union in Europe* (London, Longman).

Economist (1994), "The left in Western Europe", June 11th.

Edgren, Gösta, Karl-Olof Faxén and Clas-Erik Ohdner (1973), *Wage Formation and the Economy* (London, George Allen and Unwin).

EFTA Bulletin (1998), "After Brussels", vol. 29, no. 1, Jan.-March.

Einhorn, Eric S. and John Logue (1988), "Continuity and Change in the Scandinavian Party Systems", in Steven B. Wolinetz (ed.), *Parties and Party Systems in Liberal Democracies* (London, Routledge).

Ekbåge, Sune (1992), "Jobben, lönerna och exportberoendet", in Rolf Edberg and Ranveig Jacobsson (eds), *På tröskeln till EU* (Stockholm, Tidens förlag).

Ekström, Tord, Gunnar Myrdal and Roland Pålsson (1962), *Vi och Vasteuropa. Uppfordran till eftertanke och debatt* (Stockholm, Rabén och Sjögren).

Elgström, Ole (1990), "Socialdemokratin och det internationella solidaritet", in Bo Huldt and Klaus Misgeld (eds), *Socialdemokratin och svensk utrikespolitik. Från Branting till Palme* (Stockholm, Utrikespolitiska institutet).

Elvander, Nils (1990), "Incomes Policies in the Nordic Countries", *International Labour Review* vol. 129, no. 1, pp.1-21.

— (1994), "Självbelåten välfärdnationalism styr nej-sidan", *Svenska Dagbladet* November 6th.

Esaiasson, Peter (1996), "Kampanj på sparlåga", in Mikael Gilljam and Sören Holmberg (eds), *Ett knappt ja till EU. Väljarna och folkomröstning 1994* (Stockholm, Norstedts juridik).

Esping-Andersen, Gøsta (1985), *Politics Against Markets: The Social Democratic Road to Power* (Princeton, Princeton University Press).

— (1993), *The Three Worlds of Welfare Capitalism* (Cambridge, Polity Press).

Färm, Göran (1993), *Sverige och EG. Det nya Europa—hot eller mojlighet?* (Stockholm, Utbildningsbrevskolan).

Featherstone, Kevin (1988), *Socialist Parties and European Integration: A Comparative History* (Manchester, Manchester University Press).

Furustrand, Reynoldh and Pär Axel Sahlberg (1998), "'Regeringen är för defensiv'", *Dagens Nyheter* October 21st.

Gabel, Matthew and Harvey D. Palmer (1995), "Understanding Variation in Public Support for European Integration", *European Journal of Political Research* vol. 27, pp.3-19.

Gaffney, John (1996), "Introduction: Political Parties and the European Union", in John Gaffney (ed.), *Political Parties and the European Union* (London, Routledge).

George, Stephen (1996), *Politics and Policy in the European Union*, 3rd ed. (Oxford, Oxford University Press).

Giavazzi, Francesco (1996), "EMU—the key to the crucial German question", *Independent* May 2nd.

Gidlund, Gullan (1992), "From Popular Movement to Political Party: Development of the Social Democratic Labor Party Organization", in Klaus Misgeld, Karl Molin and Klas Åmark (eds), *Creating Social Democracy: A Century of the Social Democratic Labor Party in Sweden* (Pennsylvania, Pennsylvania Universiy Press).

— (1992), *Partiernas Europa* (Stockholm, Natur och Kultur).

Gilljam, Mikael (1996), "Den direkta demokratin", in Mikael Gilljam and Sören Holmberg (eds), *Ett knappt ja till EU. Väljarna och folkomröstning 1994* (Stockholm, Norstedts juridik).

Gradin, Anita (1989), "West European Integration: The Swedish View", *European Access* vol. 3, no. 2, p.22.

Gröning, Lotta (ed.) (1993), *Det nya riket? 24 kritiska röster om Europa-Unionen* (Stockholm, Tidens förlag).

— (1993), "En nationalism som är internationell!", in Lotta Gröning (ed.), *Det nya riket? 24 kritiska röster om Europa-Unionen* (Stockholm, Tidens förlag).

Gros, Daniel and Niels Thygesen (1992), *European Monetary Integration: From the European Monetary System to Monetary Union* (London, Longman).

Gustavsson, Jakob (1998), *The Politics of Foreign Policy Change: Explaining the Swedish Reorientation on EC Membership* (Lund, Lund University Press).

Haahr, Jens Henrik (1993), *Looking to Europe: The EC Policies of the British Labour Party and the Danish Social Democrats* (Aarhus, Aarhus University Press).

Halvarson, Arne (1995), *Sveriges statsskick: fakta och perspektiv*, 10th ed. (Stockholm, Almqvist & Wicksell).

Hamilton, Carl (1987), "Protectionism and European Economic Integration", *EFTA Bulletin* vol. 28, no. 4, pp.8-9.

Hamilton, Carl B. (1990), "'En omoralisk Europapolitik'", *Dagens Nyheter* May 31st.

Hansen, Tore and Tor Bjørklund (1996), "The Narrow Escape—Norway's No to the European Union", paper presented to European Consortium For Political joint sessions of workshops, Oslo.

Harmel, Robert and Kenneth Janda (1994), "An Integrated Theory of Party Goals and Party Change", *Journal of Theoretical Politics* vol. 6, no. 3, pp.259-87.

Hinnfors, Jonas and Jon Pierre (1998), "Currency Crises in Sweden: Policy Choice in a Globalised Economy", *West European Politics* vol. 21, no. 3, pp.103-119.

Hökmark, Gunnar, Torbjörn Pettersson and Sven-Gunnar Persson (1998), "S måste avstå från LO-miljonerna", *Dagens Nyheter* August 9th.

Hopkin, Jonathan (1996), "Parties and the Business Firm Model of Party Organisation: Cases from Spain and Italy", paper presented to Political Studies Association annual conference, Glasgow University.

Huldt, Bo (1990), "Socialdemokratin och säkerhetspolitiken", in Bo Huldt and Klaus Misgeld (eds), *Socialdemokratin och den svenska utrikespolitiken* (Stockholm, Swedish Institute of International Affairs).

Ingebritsen, Christine (1995), "Norwegian Political Economy and European Integration: Agricultural Power, Policy Legacies and EU Membership", *Conflict and Cooperation* vol. 30, no. 4, 1995, pp.349-63.

— (1998), *The Nordic States and European Unity* (Ithaca, NY, Cornell University Press).

Inglehart, Ronald (1967), *The Silent Revolution: Changing Values and Political Styles among Western Publics* (Princeton, NJ, Princeton University Press).

Jahn, Detlef and Ann-Sofie Storsved (1995), "Legitimacy through Referendum? The Nearly Successful Domino-Strategy of the EU-Referendums in Austria, Finland, Sweden and Norway", *West European Politics* vol. 18, no. 4, pp.18-37.

Jerneck, Magnus (1993), "Sweden—The Reluctant European", in Teija Tiilikainen and Ib Damgaard Petersen (eds), *The Nordic Countries and the EC* (Copenhagen, Copenhagen Political Studies Press).

Johansson, Olof, Pär Granstedt and Per-Ola Eriksson (1990), "'Ni tar inte ansvar för Sverige'", *Dagens Nyheter* June 10th.

Johansson, Sten (1993), "Löntagarstrategier inför kapitalets internationalisering", in Lotta Gröning (ed.), *Det nya riket? 24 kritiska röster om Europa-Unionen* (Stockholm, Tidens förlag).

— (1997), "'Ekonomisk galenskap, Persson!'", *Dagens Nyheter* January 19th.

— and Maj Britt Theorin (1994), "Strategigrupp utreder kärnvapens roll i EU", *Svenska Dagbladet* November 7th.

Johnson, Anders (1998), *Vi står på vägen* (Stockholm, Timbro).

Junibevægelsen (1994), *Programme of the JuneMovement* (Copenhagen, Junibevægelsen).

Katz, Richard S. and Peter Mair (1995), "Changing Models of Party Organization and Party Democracy: The Emergence of the Cartel Party", *Party Politics* vol. 1, no. 1, pp.5-28.

Kite, Cynthia (1996), *Scandinavia Faces EU: Debates and Decisions on Membership 1961-94* (Umeå, Department of Political Science, Umeå University).

Kitschelt, Herbert (1994), "Austrian and Swedish Social Democrats in Crisis: Party Strategy and Organization in Corporatist Regimes", *Comparative Political Studies* vol. 27, no. 1, pp.3-39.

— (1994), *The Transformation of European Social Democracy* (Cambridge, Cambridge University Press).

Klein, Helle (1992), "Demokrati—en fråga om inflytande och tillgänglighet", in Rolf Edberg and Ranveig Jacobsson (eds), *På tröskeln till EU* (Stockholm, Tidens förlag).

Klevenås, Lena (1993), "Den katolska kyrkan och EG", in Lotta Gröning (ed.), *Det nya riket? 24 kritiska röster om Europa-Unionen* (Stockholm, Tidens förlag).

Koeble, Thomas A. (1992), "Recasting Social Democracy in Europe: A Nested Games Explanation of Strategic Adjustment in Political Parties", *Politics & Society* vol. 20, no. 1, pp.51-70.

Kokko, Ari (1994), "Sverige: EU-medlemskapets effekter på investeringar och tillväxt", in Magnus Blomström and Robert E. Lipsey (eds), *Norden i EU: vad säger ekonomerna om effekterna?* (Stockholm, SNS förlag).

Korpi, Walter (1993), "Medlemskap = Massarbetslöshet", in Lotta Gröning (ed.), *Det nya riket? 24 kritiska röster om Europa-Unionen* (Stockholm, Tidens förlag).

Lane, Jan-Erik and Svante Ersson (1997), *Comparative Political Economy: A Developmental Approach*, 2nd ed. (London, Pinter).

— (1997), "Parties and Voters: What Creates the Ties?", *Scandinavian Political Studies* vol. 20, no. 2, pp.179-96.

Lawler, Peter (1997), "Scandinavian Exceptionalism and European Union", *Journal of Common Market Studies* vol. 35, no. 4, pp.565-94.

Lindahl, Rutger (1996), "En folkomröstning i fredens tecken", in Mikael Gilljam and Sören Holmberg (eds), *Ett knappt ja till EU. Väljarna och folkomröstning 1994* (Stockholm, Norstedts juridik).

Lindbeck, Assar (1975) *Swedish Economic Policy* (London, Macmillan).

— , Per Molander, Torsten Persson, Olof Petersson, Agnar Sandmo, Birgitta Swedenborg and Niels Thygesen (1994), *Turning Sweden Around* (Cambridge, Mass., MIT Press).

Lindblom, Stig (1967), "Bollen ligger inte hos oss: EEC och anslutningsformerna", *Tiden* vol. 59, pp.589-601.

Lindström, Ulf (1994), *Euro-Consent, Euro-Contract or Euro-Coercion? Scandinavian Social Democracy, the European Impasse and the Abolition of Things Political* (Oslo, Scandinavian University Press).

Lipset, Seymour Martin and Stein Rokkan (1967), "Cleavage Structures, Party Systems, and Voter Alignment", reprinted in Peter Mair (ed.), *The West European Party System* (Oxford, Oxford University Press, 1991).

Löwdwin, Per (1998), *Det dukade bordet. Om partierna och de ekonomiska kriserna* (Uppsala, Acta Universitatis Upsaliensis).

Luif, Paul (1990), "Austria's Application for EC Membership: Historical Background, Reasons and Possible Results", in Finn Laursen (ed.), *EFTA and the EC: Implications of 1992* (Maastricht, European Institute of Public Administration).

Lundberg, Erik (1985), "The Rise and Fall of the Swedish Model", *Journal of Economic Literature* vol. 23, pp.1-36.

Mair, Peter (1991), "Introduction", in Peter Mair (ed.), *The West European Party System* (Oxford, Oxford University Press).

— (1994), "Party Organizations: From Civil Society to the State", in Richard S. Katz and Peter Mair, *How Parties Organize: Change and Adaptation in Party Organizations in Western Democracies* (London, Sage).

Mattli, Walter (1996), "Regional Integration and the Enlargement Issue: A Macroanalysis", in Gerald Schneider, Patricia A. Weitsman and Thomas Bernauer (eds), *Towards a New Europe: Stops and Starts in Regional Integration* (Westport, Cn., Praeger).

Martin, Andrew (1996), "Macroeconomic Policy, Politics, and the Demise of Central Wage Negotiations in Sweden", Centre for European Studies Working Paper Series 63 (Cambridge, Mass., Harvard University).

Meidner, Rudolf (1993), "Neutralitet och fullsysselsättning omodernt i ett EG-anslutet Sverige", in Lotta Gröning (ed.), *Det nya riket? 24 kritiska röster om Europa-Unionen* (Stockholm, Tidens förlag).

— (1994), "The Rise and Fall of the Swedish Model", in Wallace Clement and Rianne Mahon (eds), *Swedish Social Democracy: A Model in Transition* (Toronto, Canadian Scholars' Press).

Micheletti, Michele (1995), *Civil Society and State Relations in Sweden* (Aldershot, Avebury).

Miljan, Toivo (1977), *The Reluctant Europeans: The Attitudes of the Nordic Countries towards European Integration* (London, C. Hurst and Co.).

Milner, Henry (1989), *Sweden: Social Democracy in Practice* (Oxford, Oxford University Press).

Ministry of Finance (Sweden) (1998), "EMU Facts" (www.regeringen.se).

Misgeld, Klaus (1990), "Den svenska socialdemokratin och Europa—från slutet av 1920-talet till början av 1970-talet. Attityder och synsätt i centrala uttalande och dokument", in Bo Huldt and Klaus Misgeld (eds), *Socialdemokratin och den svenska utrikespolitiken* (Stockholm, Swedish Institute of International Affairs).

Moravcsik, Andrew (1993), "Preferences and Power in the European Community: A Liberal Intergovernmentalist Approach", *Journal of Common Market Studies* vol. 31, no. 4, pp.473-521.

Mortimer, Edward (1993), "Same deal as before: the Danes did not win new concessions on Maastricht", *Financial Times* January 27th.

Mouritzen, Hans (1988), *Finlandization: Towards a General Theory of Adaptive Politics* (Aldershot, Avebury).

Olson, Mancur (1990), *How Bright are the Northern Lights? Some Questions About Sweden* (Lund, Lund University Press).

Oscarsson, Henrik (1996), "EU-dimensionen", in Mikael Gilljam and Sören Holmberg (eds), *Ett knappt ja till EU. Väljarna och folkomröstning 1994* (Stockholm, Norstedts juridik).

Oskarson, Maria (1996), "Väljarnas vågskålar", in Mikael Gilljam and Sören Holmberg (eds), *Ett knappt ja till EU. Väljarna och folkomröstning 1994* (Stockholm, Norstedts juridik).

Padoa-Schioppa, Tommaso et al (1987), *Efficiency, Stability and Equity: A Strategy for the Evolution of the Economic System of the European Community* (Oxford, Oxford University Press).

Pedersen, Thomas (1994), *European Union and the EFTA Countries: Enlargement and Integration* (London, Pinter).

Petersen, Nikolaj (1993), "'Game, Set and Match for Denmark': Denmark and the EU After Edinburgh", in Teija Tiilikainen and Ib Damgaard Petersen, *The Nordic Countries and the European Community* (Copenhagen, Copenhagen Political Studies Press).

Petersson, Olof (1994), *The Government and Politics of the Nordic Countries*, translated by Frank Gabriel Perry (Stockholm, Fritzes).

Pettersson, Kenth, Bengt-Ola Ryttar, Lena Sandlin and Sören Wibe (1996), "'Vi stoppar EMU-inträdet'", *Dagens Nyheter* May 19th.

Pettersson, Lars (1992), "Sweden", in David A. Dyker (ed.), *The National Economies of Europe* (London, Longman).

Pierre, Jon and Anders Widfeldt (1994), "Party Organizations in Sweden: Colossuses with Feet of Clay or Flexible Pillars of Government?", in Richard S. Katz and Peter Mair, *How Parties Organize: Change and Adaptation in Party Organizations in Western Democracies* (London, Sage).

Pontusson, Jonas (1992), *The Limits of Social Democracy: Investment Politics in Sweden* (Ithaca, NY, Cornell University Press).

— (1994), "Sweden: After the Golden Age", in Perry Anderson and Patrick Camiller (eds), *Mapping the West European Left* (London, Verso).

— (1995), review of Kitschelt, *The Transformation of European Social Democracy* (1994), *Comparative Political Studies* vol. 28, no. 3, pp.469-75.

Przeworski, Adam and John Sprague (1986), *Paper Stones: A History of Electoral Socialism* (Chicago, University of Chicago Press).

Ross, John F.L. (1991), "Sweden, the European Community, and the Politics of Economic Realism", *Cooperation and Conflict* vol. 26, no.3, pp.117-28.

Rothstein, Bo (1996), *The Social Democratic State: The Swedish Model and the Bureacratic Problem of Social Reforms* (Pittsburgh, PA, University of Pittsburgh Press).

Ruin, Olof (1988), "Sweden: The New Constitution (1974) and the Tradition of Constitutional Politics", in Vernon Bogdanor (ed.), *Constitutions in Democratic Politics* (Aldershot, PSI/Gower).

— (1996), "Sweden: The Referendum as an Instrument for Defusing Political Issues", in Michael Gallagher and Pier Vincenzo Luigi (eds), *The Referendum Experience in Europe* (London, Macmillan).

Ryner, Magnus (1994), "Economic Policy in the 1980s: The 'Third Way', the Swedish Model and the Transistion from Fordism to Post-Fordism", in Wallace Clement and Rianne Mahon (eds), *Swedish Social Democracy: A Model in Transition* (Toronto, Canadian Scholars' Press).

Sainsbury, Diane (1992), "The 1991 Swedish Election: Protest, Fragmentation, and a Shift to the Right", *West European Politics* vol. 15, no. 2, pp.160-66.

Sannerstedt, Anders and Mats Sjölin (1992), "Sweden: Changing Party Relations in a More Active Parliament", in Eric Damgaard (ed.), *Parliamentary Change in the Nordic Countries* (Oslo, Scandinavian University Press).

SAP (1992a), *Socialdemokratin inför EG-förhandlingarna*, Politisk Redovisning Nummer 2 (Stockholm, SAP).

— (1992b), *Verksamhetsberättelse 1990-92* (Stockholm, SAP).

Schlesinger, Joseph A. (1984), "On the Theory of Party Organisation", *The Journal of Politics* vol. 46, pp. 369-400.

Schneider, Gerald and Patricia A. Weitsman (1996), "The Punishment Trap: Integration Referendums as Popularity Contests", *Comparative Political Studies* vol. 28, no. 4, pp.582-607.

Schwok, René (1995), "EC-EFTA Relations, A Critical Assessment", paper presented to the Second Pan-European Conference in International Relations, Paris.

Sifo Research and Consulting (www.sifo.se).

Simon, Herbert A. (1985), "Human Nature in Politics: The Dialogue of Psychology with Political Science", *American Political Science Review* vol. 79, no. 2, pp.293-304.

Sjöstedt, Gunnar (1987), *Sweden's Free Trade Policy: Balancing Economic Growth and Security* (Stockholm, Swedish Institute).

Spånt, Roland (1993a), "Jobban offras för stabila priser", *LO-tidningen* January 29th.

— (1993b), "SAP—ett slag i luften!", in Lotta Gröning (ed.), *Det nya riket? 24 kritiska röster om Europa-Unionen* (Stockholm, Tidens förlag).

Stålvant, Carl-Einar (1976), "The Exit vs Voice Option: Six Cases of Swedish Participation in International Organisations", *Cooperation and Conflict* vol. 11, p.43.

— (1990), "Rather a Market than a Home, But Preferably a Home Market: Swedish Policies Facing Changes in Europe", in Finn Laursen (ed.), *EFTA and the EC: Implications of 1992* (Maastricht, European Institute of Public Administration).

— and Carl Hamilton (1991), "Sweden", in Helen Wallace (ed), *The Wider Western Europe: Reshaping the EC-EFTA Relationship* (London, Pinter).

Stenberg, Ewa and Gunnar Örn (1996), "De spelade bort sexton miljarder", *Dagens Nyheter* November 3rd 1996.

Stigandal, Mikael (1995), "The Swedish Model: Renaissance or Retrenchment?", *Renewal* vol. 3, no. 1, pp.14-23.

Statistics Sweden (www.scb.se).

Stråth, Bo (1992), *Folkhemmet mot Europa. Ett historisk perspektiv på 90-talet* (Stockholm, Tiden).

Strom, Kaare (1990), "A Behavioural Theory of Competitive Political Parties", *American Journal of Political Science* vol. 34, no. 2, pp.565-98.

Svåsand, Lars and Ulf Lindström (1996), "Scandinavian Political Parties and the European Union", in John Gaffney (ed.), *Political Parties and the European Union* (London, Routledge).

Tarschys, Daniel (1971), "Neutrality and the Common Market: The Soviet View", *Cooperation and Conflict* vol. 11, pp.65-75.

Thorsson, Inga (1962), "Sverige, svensk socialdemokrati om omvärlden", *Tiden* vol. 54, pp.262-67.

Tiden (1960), "Bonn och Europa", vol. 3, pp.181-84.

— (1961a), "Det europeiska alternativet", vol. 53, pp.313-17.

— (1961b), "Neutraliteten och De Sex", vol. 53, pp.386-88.

— (1961c), "Psychologiska förberedelser för EEC", vol. 53, pp. 449-452.

Tilton, Tim (1991), *The Political Theory of Swedish Social Democracy* (Oxford, Clarendon Press).

Tingsten (1973), Herbert, *The Swedish Social Democrats: Their Ideological Development* (New Jersey, Bedminster Press).

Tsebelis, George (1990), *Nested Games: Rational Choice in Comparative Politics* (Los Angeles, University of California Press).

Tulard, Jean (1984), *Napoleon: The Myth of the Saviour* (London, Methuen and Co.).

Uusmann, Ines (1992), "Kvinnorna är halva EU—svenska kvinnors liv i morgondagens Europa", in Rolf Edberg and Ranveig Jacobsson (eds), *På tröskeln till EU* (Stockholm, Tidens förlag).

Valen, Henry (1973), "Norway: 'No' to EEC", *Scandinavian Political Studies* vol. 8, pp.214-26.

Vaubel, Roland (1995), *The Centralisation of Western Europe* (London, Institute of Economic Affairs).

Vogel, Hans (1983), "Small States' Efforts in International Relations: Enlarging the Scope", in Ottmar Höll (ed.), *Small States in Europe and Dependence* (Vienna, Austrian Institute for International Affairs).

Voronov, Konstanti (1990), "Sovjet välkommer svensk EG-anslutning", *Dagens Nyheter* June 30th.

Wahlbäck, Krister (1986), *The Roots of Swedish Neutrality* (Stockholm, Swedish Institute).

Wallensteen, Peter (1990), "Socialdemokratin och säkerhetspolitik: några kommentarer", in Bo Huldt and Klaus Misgeld (eds), *Socialdemokratin och svensk utrikespolitik. Från Branting till Palme* (Stockholm, Utrikespolitiska institutet).

Ware, Alan (1996), *Political Parties and Party Systems* (Oxford, Oxford University Press).

Westerberg, Bengt (1990), "'Patetiskt, Olof Johansson!'", *Dagens Nyheter* June 12th.

Wibe, Sören (1993), "EG och ekonomin", in Lotta Gröning (ed.), *Det nya riket? 24 kritiska röster om Europa-Unionen* (Stockholm, Tidens förlag).

Widfeldt, Anders (1997), *Linking Parties with People? Party Membership in Sweden 1960-1994* (Gothenburg, Department of Political Science, Göteborg University).

Winberg, Margareta (1993), "Dyrt, byråkratiskt, miljömassigt sämre—om EGs jordbrukspolitik!", in Lotta Gröning (ed.), *Det nya riket? 24 kritiska röster om Europa-Unionen* (Stockholm, Tidens förlag).

Worre, Torben (1987), "The Danish Euro-Party System", *Scandinavian Political Studies* vol. 10, no. 1, pp.79-95.

— (1988), "Denmark at the Crossroads: The Danish Referendum of 28 February 1986 on the EC Reform Package", *Journal of Common Market Studies* vol. 26, no.4, pp.361-88.

— (1995), "First No, then Yes: The Danish Referendums on the Maastricht Treaty 1992 and 1993", *Journal of Common Market Studies* vol. 33, no. 2, pp.235-57.

Index

Printed and bound by CPI Group (UK) Ltd, Croydon, CR0 4YY

21/10/2024

01777087-0005